SOCIETY FOR EXPERIMENTAL BIOLOGY
SEMINAR SERIES · 13

ANIMAL MIGRATION

ANIMAL MIGRATION

Edited by

D. J. AIDLEY

School of Biological Sciences, University of East Anglia, Norwich

CAMBRIDGE UNIVERSITY PRESS

Cambridge
London   New York   New Rochelle
Melbourne   Sydney

Published by the Press Syndicate of the University of Cambridge
The Pitt Building, Trumpington Street, Cambridge CB2 1RP
32 East 57th Street, New York, NY 10022, USA
296 Beaconsfield Parade, Middle Park, Melbourne 3206, Australia

First published 1981

Printed in USA by
Vail-Ballou Press, Inc.,
Binghamton, N.Y.

Library of Congress catalogue card number: 81-3905

British Library Cataloguing in Publication Data

Animal migration. – (Society for Experimental
Biology seminar series; 13)

1. Animal migration – Congresses
I. Aidley, D. J.   II. Series
591.52   QL754

ISBN 0 521 23274 0 hard covers
ISBN 0 521 29888 1 paperback

CONTENTS

CONTRIBUTORS

Aidley, D. J.
School of Biological Sciences, University of East Anglia, Norwich, NR4 7TJ, UK.
Alerstam, T.
Department of Animal Ecology, University of Lund, S-223 62 Lund, Sweden.
Arnold, G. P.
Fisheries Laboratory, Lowestoft, Suffolk, NR33 0HT, UK.
Baker, R. R.
Department of Zoology, Williamson Building, University of Manchester, Manchester, M13 9PL, UK.
Brown, S. G.
Sea Mammal Research Unit, c/o British Antarctic Survey, Madingley Road, Cambridge, CB3 0ET, UK.
Harden Jones, F. R.
Fisheries Laboratory, Lowestoft, Suffolk, NR33 0HT, UK.
Joyce, R. J. V.
College of Aeronautics, Cranfield Institute of Technology, Cranfield, Bedford, MK43 0AL, UK.
Keeton, W. T.
Langmuir Laboratory, Cornell University, Ithaca, New York 14850, USA
Lockyer, C. H.
Sea Mammal Research Unit, c/o British Antarctic Survey, Madingley Road, Cambridge, CB3 0ET, UK.
Matthews, G. V. T.
The Wildfowl Trust, Slimbridge, Gloucester, GL2 7BT, UK.
O'Connor, R. J.
British Trust for Ornithology, Beech Grove, Tring, Hertfordshire, HP23 5NR, UK.
Southwood, T. R. E.
Department of Zoology, University of Oxford, Oxford, OX1 3PS, UK.

PREFACE

In recent years there seems to have been an upsurge of interest in the migratory behaviour of animals. Twenty years ago, to take some examples, night-flying insects were not being tracked by radar, the sensitivity of birds to magnetic fields was generally thought to be mythical, and there was little discussion as to the selected advantages of migration. Things are different today.

The Society for Experimental Biology held a Seminar Series symposium on the subject of Animal Migration at its Lancaster meeting in December 1979. This book represents the proceedings of that session. The contributors were asked to describe their own particular interests in a manner suitable for other biologists; I hope that the resulting collection will serve to illustrate the range of modern studies on the subject.

As convenor of the symposium and editor of the book I am very grateful for the help and advice which I have received from a number of people, including especially Dr F. R. Harden Jones and Dr G. Shelton. The project probably evolved from an undergraduate course on migration and orientation in animals, run as a collaborative venture by the Fisheries Laboratory, Lowestoft, and the University of East Anglia.

It was very saddening to learn recently of the death of one of the contributors to this book, Professor William T. Keeton. His account of our knowledge of the mechanisms of bird orientation and navigation, delivered with elegance, clarity and wit, was one of the highlights of the Lancaster meeting. His contributions to knowledge in this field will remain as a lasting memorial to his scientific acumen.

December 1980   *David Aidley*
                              Editor for the Society for Experimental Biology

D.J.AIDLEY

Questions about migration

The study of animal migration involves a variety of different approaches. The purpose of this book is to illustrate how some of these techniques or ways of thinking have been applied to different groups of animals. But first, as an overture to what follows, it may be useful to consider briefly some of the questions which may be raised by the study of animal migration, both as a general phenomenon and in individual instances.

What do we mean by migration?

There are difficulties in providing a generally agreed definition of migration because different people, and especially students of different animal groups, use the term in different ways. In birds, for example, the long-range seasonal movements of the European swallow or the white stork are universally regarded as migration. Thus Landsborough Thomson (1964) defines migration as a 'regular movement of birds between alternate areas inhabited by them at different times of year, one area being that in which the birds breed and the other being in an area better suited to support them at the opposite season'. This definition probably includes the movements of dispersal away from the breeding colonies that occur in many seabirds after breeding, but excludes the occasional irruptive movements of crossbills and the nomadic movements of queleas: and are the diurnal movements of such birds as starlings and geese between their roosts and their feeding grounds a form of migration? Yet all these different types of movement have some features in common: they occur in relation to the birds' needs, for food, shelter or breeding sites. Furthermore the conceptual boundaries between different types of movement may be indistinct and subject to considerable overlap.

When we come to look at other animal groups the concepts used in bird migration studies may not be readily applicable. The mass movements of aphids and locusts, for example, are referred to as migrations, yet there is little expectation of the return of the population (and certainly not of the individual, whose life cycle is shorter than the seasonal cycle) to the particular place from which it departed. The term migration is also applied to certain circadian movements such as the vertical migration of planktonic crustaceans and the nightly migration of parasitic filarial worms from deep tissues to just under the skin (a movement which enables them to be transferred to night-biting mosquitoes).

When considering animal migration as a whole, then, there are many reasons for adopting a broad inclusive definition rather than a narrow exclusive one. The Oxford English Dictionary, for example, defines migration generally as 'the action of moving from one place to another'. (A more restrictive definition is given later, however: 'of animals: the action of moving in flocks, shoals, etc. from one region or habitat to another; especially of some birds and fishes, the periodical departure from and return to a region at a particular season of the year'.) Baker (1978) adopts a similarly broad definition as 'the act of moving from one spatial unit to another'. I would prefer to follow Southwood (1962) and others in regarding migration as a movement from one habitat to another, a view which emphasizes the fact that the animal in some way changes the conditions of its life by undertaking migration.

All this may seem rather unsatisfactory to those in search of precise, clear-cut, exclusive definitions. But such definitions are of use only in considering ecologically and taxonomically restricted groups of animals. Faced with the grand complexity of the living world, we must recognize the limitations of our language and our concepts.

What are the movements of migrants?

Perhaps the first question to be answered in regarding any particular population of migrants is to determine where it goes to and comes from on its migratory journeys. There are various ways of tackling this problem.

Changes in the population density of a species often give clues to its migration. The occurrence of fieldfares in Britain in winter suggests that they have migrated there from their breeding grounds in northern Europe. The movements of commercial fish have been estimated from the seasonal variations in the catch at different sites.

Sometimes migration can be inferred from sequential changes in the infraspecific characters of the members of a species sampled at a single place. Different wader populations have been distinguished on the basis of bill length and body colour. An influx of armyworm moths at a site in Tanzania has been deduced from a sudden change in the wing length and sex ratio of moths sampled at a light trap (Aidley & Lubega, 1979).

Animal movements can sometimes be observed visually, as for example in swallows, storks, aphids and grey whales. Nocturnal bird movements have been observed by moon-watching. Radar has been used to track the movements of birds and insects, and sonar has been used to track fish.

The movements of individuals have often been investigated by marking them in some way and then recapturing them later at some distant site. Thus fish have had a fin removed, birds have been dyed, and various insects have been dyed, made radioactive, or fed on rubidium as larvae. These methods label the members of a population but they do not label them individually. Effective individual labelling was begun by Mortensen, who in 1899 labelled 164 starlings in Denmark with numbered rings which also gave an address for the ring to be returned to. Bird ringing schemes are now worldwide. The British Trust for Ornithology's scheme, to take an example, rings about half-a-million birds each year, and receives about 12 000 recoveries (Mead, 1974). Individual tagging systems have also been used for bats, fish, whales, seals, turtles, frogs and the monarch butterfly.

The results of these endeavours are often fascinating and sometimes dramatic. Swallows from different parts of Europe winter in different parts of Africa. Wigeon wintering in Britain may return to Siberia in the spring. Young grey seals disperse from the island of North Rona to the coasts of Scotland, Norway, the Faeroes, Iceland and Ireland. Individual monarch butterflies may fly over 2000 miles from Canada to Mexico.

What factors initiate migration?

Many migrations are annual events, and hence their timing is in relation to the annual cycle. A possible cue is the cyclical change in day length that occurs in non-tropical regions. Migratory behaviour (often measured as an increase in nocturnal restlessness) has been initiated in winter in a number of birds by increasing their day lengths (see Berthold, 1975). Day-length changes cannot act as a cue for migrants wintering in equatorial regions, however. Here there is evidence that there are endogenous circannual rhythms in some birds; willow warblers and garden warblers showed nocturnal restlessness at the appropriate times when kept for up to 33 months at constant day length (Gwinner, 1977). It seems likely that hormonal changes are involved in the links between the timing event and the change in behaviour.

Factors initiating less regular movements are likely to be environmental. The 'irruptions' (irregular mass movements out of their normal range) of crossbills and some other northern forest birds seem to be related to the size of the seed crop.

Many insect migrations are also related to environmental changes. Crowding may also be important here, and may act at an earlier stage in the developmental process so as to produce specifically migratory individuals. In locusts, for example, crowding produces the gregarious phase insects which are more active than those in the solitary phase. In some aphids crowding or a reduction in the nutrient value of the host plant produces winged individuals from a previously wingless population (see Dixon, 1973).

Once the animal is in the right condition to migrate, it may need some further environmental cue to initiate the actual migratory movement. A bird may wait for the right weather conditions (or at least for the disappearance of the wrong ones, such as fog or very strong winds), a fish may wait for the right tide, and an aphid may have a particular temperature threshold for flight.

How do migrants find their way?

In order to find its goal at a different position on the earth's surface an animal must possess certain sensory systems and have certain decision-making programmes in its behaviour. A migrating aphid, for example, needs to have the sensory equipment to detect a suitable habitat (which may be at least partly defined by the fact that it reflects yellow rather than blue light) and the behavioural programme to descend then from the airstream in which it is travelling. A European swallow migrating in autumn must have the ability to pick a southerly direction and some means of determining its winter quarters when it arrives there.

The ability to take up a particular direction with respect to some feature or property of the environment is called *orientation*. The orienting animal is rather like a man who possesses a compass and an instruction to proceed in some particular direction. Animals may orient with respect to objects on the earth's surface (including such possibilities as coastal profiles for grey whales and thunderstorms for wildebeest), to the sun, to the stars, to the earth's magnetic field, to odour sources, to directions of current flow, and so on. Furthermore, they may use more than one of these environmental sources of information at any one time, and may use different ones at different times. This leads to complications in experimental design: if an animal shows evidence of adequate orientation in the absence of some particular type of sensory information, we cannot conclude that it does not normally use that information.

The term *navigation* usually implies rather more than orientation. It is as if we had given our man with the compass a map which enables him to see where he is and where he wants to go. Schmidt–Koenig (1979) defines navigation as the capacity to find or establish reference to a goal. Most human technologically based navigation systems are bicoordinate: they establish positions by finding their coordinates of latitude and longitude, for example, ultimately employing astronomical observations to do so.

Homing experiments have clearly shown that many animals are capable of some form of navigation, but attempts to determine the mechanisms by which the animal establishes the relative positions of its starting point and its goal have been much less successful. The nature of the map turns out to be much more elusive than the nature of the compass. One possibility, emphasized by Baker (1978), is that animals may build up knowledge of a 'familiar area' as a result of exploratory activity, especially in the immediate post-juvenile phase. The 'map' is then based on the animal's personal experience, supplemented by knowledge of what can be detected outside the familiar area by the use of the distance senses such as vision and hearing. But many of the results of homing experiments do not seem to be explicable in this way.

How much energy is used in migration?

The energy costs of covering a particular distance differ with the type of locomotion used. A typical 100-g mammal might consume 2400 J in running 1 km (the figures are estimated from data given by Schmidt–Nielsen, 1972), a 100-g bird might consume 680 J in flying 1 km, and a 100-g fish might consume 240 J in swimming 1 km. Hence 1 g of fat used as the fuel (and supplying 37 kJ) would take the mammal 15 km, the bird 54 km and the fish 154 km.

Many migratory birds put on considerable amounts of fat before take-off on long distance flights across seas or deserts; in some small passerines the fat fuel store may be as much as 50% of the body weight. If our 100-g bird had 30 g of fat, it would have a range of 1630 km, just sufficient to cross the western Sahara.

Migrants will not need to accumulate fuel reserves before a journey if they are able to feed while on the journey. This option is presumably open to many fish, grazing mammals, and aerial feeders such as swallows. It is noteworthy that swallows, unlike most small passerines, migrate by day (when they are able to feed) and do not accumulate large amounts of fat.

Sometimes energy for migration is partly supplied by the environment in the form of air or water currents. Soaring birds utilize up-currents, aphids and other small insects ride on the winds, and many fish utilize tidal or oceanic currents.

When considering the net energetic cost of migration, it is also necessary to consider what the animal would be doing if it were not migrating. I suspect that swifts, for example, which spend most of their lives on the wing, use much the same amount of energy whether or not they are migrating. Similar considerations may apply to fish and whales. But the migratory flights of such birds as warblers, who would otherwise be roosting, must constitute a very great increase in total energy expenditure.

What is the selective advantage of migration?

Modern biologists explain the existence of particular modes of animal behaviour in terms of evolution by natural selection. Migrants migrate because their migrating ancestors left more descendants than did their conspecifics which did not migrate. But what circumstances would lead to such reproductive success?

Migration allows migrants to exploit the resources of habitats which are not permanently suitable for their mode of life. The arctic summer, for example, with its long day length and brief abundance of invertebrate life, is highly suited for the breeding of a variety of wading and other birds. Such birds would be quite unable to survive the arctic winter and therefore have to be medium- or long-distance migrants. (The alternative strategy, utilized by arctic insects and some mammals, is dormancy, either as diapausing eggs or pupae or by hibernation.)

Another example is provided by the swallows and martins (Hirundinidae). There are 31 species of swallows resident in subsaharan Africa (although many of them migrate within the continent), two species which migrate there from the Mediterranean region, and three (the European swallow, the house martin and the sand martin) which migrate there from almost the whole of the western palaearctic region. Since the aerial insect population in the north temperate winter is negligible, it is reasonable to suppose that these three species evolved from populations which originally bred in Africa. By migrating northwards for breeding they would have found regions where there was less competition for flying insects, more daylight hours in which to catch them, and there were fewer predators of eggs and nestlings. The consequence of these factors is seen in the success of the migrant hirundines: Moreau (1972) estimated that there are about 90 000 000 house martins, 220 000 000 swallows and 375 000 000 sand martins setting out for Africa each autumn. I suspect that these numbers are appreciably greater than those of most if not all of the resident African species of the family.

When environmental resources vary in a less-predictable manner than do those linked to the seasonal changes in the north temperate zone, we may expect to see less regular migrations. The migrations of queleas and locusts are more varied from year to year than are those of palaearctic swallows. Nevertheless the ability to migrate provides the possibility of exploiting food resources whose occurrence is irregular and temporary, provided some method of locating those resources exists.

Migrants and men

An anthropocentric view divides migrants into good ones, whose populations we wish to conserve or increase, and bad ones, whose populations we wish to reduce. Most people would regard swallows, for example, as good migrants (they make Europeans more cheerful and eat some of their aphids) and locusts as bad ones. Occasionally the categories overlap, as when the black-tailed godwits for whom (with other waders) reserves are provided in western Europe eat newly germinated rice seedlings in west Africa.

Effective conservation and effective control both demand a good knowledge of the biology of the species involved. With migratory animals this may be quite difficult, since vital factors in the animals' lives may exert their influence in some distant region of which the would-be manager has little knowledge and over which he has no control.

Migratory animals can usually cross political frontiers with ease, whereas human activities are greatly affected by them. Thus we find that many of the birds which are legally protected in northern Europe are shot for pleasure or profit in southern Europe. Countries which have closed their borders or are at war with each other, as has happened in eastern and northeastern Africa, may find it difficult or impossible to coordinate their pest control programmes. Nevertheless, given time, good sense and hard work, it may be possible to overcome these problems. In this overcrowded world it is very necessary that we should do so.

The need for management provides a direct reason for the study of migration. Another is that migration involves problems of fundamental importance in zoology. And it is possible – as Baker shows later in this book – that knowledge of animal migration can help us to understand certain aspects of human life. But perhaps the main reason for the interest of zoologists in migration is less logical but more pervasive. Migrants are often beautiful, they may journey great distances to faraway places, they act as though they were adventurous, intrepid, free, as though they solved their problems by taking action. They stir the imagination.

References and further reading

Aidley, D. J. & Lubega, M. (1979). Variation in wing length of the African Armyworm, *Spodoptera exempta,* in east Africa during 1973–74. *Journal of Applied Ecology,* **16,** 653–62.

Baker, R. R. (1978). *The Evolutionary Ecology of Animal Migration.* London: Hodder and Stoughton.

Berthold, P. (1975). Migration: control and metabolic physiology. In *Avian Biology,* vol. **5,** ed. D. S. Farner & J. R. King, pp. 77–128. New York: Academic Press.

Dixon, A. F. G. (1973). *Biology of Aphids.* London: Edward Arnold.

Gunn, D. L. & Rainey, R. C. (1979). *Strategy and Tactics of Control of Migrant Pests.* London: The Royal Society. (Also published in *Philosophical Transactions of the Royal Society of London,* B **287,** 245–488.)

Gwinner, E. (1977). Circannual rhythms in bird migration. *Annual Review of Ecology and Systematics,* **8,** 381–405.

Harden Jones, F. R. (1968). *Fish Migration.* London: Edward Arnold.

Johnson, C. G. (1969). *Migration and Dispersal of Insects by Flight.* London: Methuen & Co.

Mead, C. (1974). *Bird Ringing.* Tring: British Trust for Ornithology.

Moreau, R. E. (1972). *The Palaearctic–African Bird Migration Systems.* London: Academic Press.

Schmidt–Koenig, K. (1975). *Migration and Homing in Animals.* Berlin: Springer–Verlag.

Schmidt–Koenig, K. (1979). *Avian Orientation and Navigation.* London: Academic Press.

Schmidt–Nielsen, K. (1972). Locomotion: energy cost of swimming, flying and running. *Science,* **177,** 222–8.

Southwood, T. R. E. (1962). Migration of terrestrial arthropods in relation to habitat. *Biological Reviews,* **37,** 171–214.

Thomson, A. Landsborough (1964). Migration. In *A New Dictionary of Birds,* pp. 465–72, London: Nelson.

THOMAS ALERSTAM

The course and timing of bird migration

Think of yourself as pilot of a light aircraft confronted with the enterprise of completing, in one or two months, a journey lasting 50 to 200 hours of continuous flight. Imagine also that winds on average blow about five times faster than normal, so that wind speed regularly amounts to a large fraction of the speed of your aircraft, and sometimes even exceeds the aircraft's speed. Petrol is precious and in variable supply at various prices along your flight route. Your task is to complete the journey with minimum costs without exposing yourself to undue hazards, and without getting delayed.

You would indeed have a lot to calculate before taking off: What is the most economical flight speed of your aircraft, and how does it vary with winds, flight altitude, and the extra weight of fuel reserves? Perhaps the gliding performance of your aircraft is good enough to permit you to travel some distances by soaring in updraughts, with the engine off. It may even be economical to make detours in order to stay over regions with strong updraughts. Would it be favourable to bring along large fuel reserves and fly non-stop for long distances without refuelling, or to travel by numerous short flights thereby saving transport costs for extra fuel? To solve that problem, you have to know about possible refuelling stations along your route, about petrol prices, landing fees, and the time delay you will incur by landing at a particular site.

Which is the optimal route to your destination? Perhaps it pays to follow the great circle, or you should make a detour where you can benefit from favourable winds. Of course you will fly only on days when the weather is as favourable as possible, but which is the best weather, and how choosy can you afford to be? What flight altitude will you prefer under different weather conditions? You must also consider whether and how you will compensate for wind drift, when to fly low along coast-lines and other leading-lines to gain protection from the wind and avoid drift, and how to exploit the winds at high altitudes. Questions abound and you would have to spend a long time looking up facts and making calculations before you arrived at a reasonable strategy for your flight.

Your situation is analogous to that of a migrating bird. The migratory strategy of a bird, a harmonious mixture of rigid and flexible behaviour adapted to a bewildering number of factors affecting the safety and economics of the migratory journey, has been worked out by natural selection. Selection goes on to maintain and refine the level of adaptation in the birds' migratory strategies.

In this chapter I will give some glimpses of different solutions that migrating birds have found to the above kinds of problem.

Flight speed

What are the optimal flight speeds in migrating birds? To answer this question one must look into the theory of bird flight mechanics. For this presentation I will use the model developed by Pennycuick (1969, 1975). Related versions of this model have been presented by Tucker (1973) and Greenewalt (1975). Quite recently, Rayner (1979) tried a new approach to this problem by developing a model on the basis of vortex theory applied to the motion of air in the wake of a flying bird.

A bird in horizontal flight expends power to support its weight (this power component is called induced power), to overcome the drag of the body (parasite power) and of the flapping wings (profile power). Furthermore, power is required for general maintenance, including basal metabolism and extra pumping of blood and ventilation of lungs. These power components can be summed up for various flight velocities (V) as shown in Fig. 1. Induced power falls approximately in proportion to V^{-1}, while the parasite power increases in proportion to V^3. Basal metabolism and profile power are assumed to be constant irrespective of flight speed. For the profile power this assumption is not theoretically justified, but studies of pigeons flying in a wind tunnel suggest it is a reasonable approximation. As a result, the equation relating total power to flight speed will be of the general form:

$$P = k + aV^{-1} + bV^3$$

where P is power, V is flight velocity and k, a and b are constants that vary due to the bird's mass, form and wing span, to air density and coefficient of drag. This power equation does not apply to extremely low flight speeds or hovering ($V = 0$). For these cases a different method of calculating power must be used.

The power curve will be U-shaped as seen in Fig. 1, and there are two specific flight velocities of particular interest. The most obvious is where total power is minimized. This velocity is called V_{mp} (mp is minimum power) and may be calculated by solving $dP/dV = 0$. However, for migrating birds the primary interest is not to minimize the absolute rate of energy consumption but rather to minimize energy consumption per unit of distance flown. This will allow migrants to cover the maximum range on a given energy reserve. V_{mr} (mr is maximum range) is found by solving $dP/dV = P/V$, i.e. drawing a tangent from the origin to the power curve as shown in Fig. 1. It is easy to see that this gives the minimum possible ratio of P to V.

Studies on flying birds under experimental conditions have yielded empirical approximations for various parameters in the power equation, making it possible to estimate this equation on the basis of the bird's mass and wing span only. In this way I have estimated the power equation for low-altitude flight in a selected number of species of various sizes (Table 1), and on the basis of these equations I have calculated V_{mp} and V_{mr}. The estimated power curve is presented in full for the wood pigeon in Fig. 1. Note that the power given in this figure refers to power output. The corresponding power input is the rate at which the bird consumes chemical energy (the ratio of power output to input, i.e. the mechanical efficiency of muscles and supporting systems, is about 23%).

The species in Table 1 were selected because there are records available of radar observations of flight speeds of migrating flocks and individuals in these birds. From this table it is seen that the mean air speed (flight speed corrected for the influence of wind) of the migrants generally is close to V_{mr}. Hence, in their choice of flight speed migrants seem keen to minimize energy consumption per unit of distance covered. The only exception in the Table is the whooper swan, whose air speed is closer to V_{mp} than V_{mr}. One possible explanation is that air speed given in the Table is not typical for migrating swans as it is based

Fig. 1. Power versus speed for horizontal bird flight. Total power is the sum of parasite, induced, profile, and maintenance power as shown in (a). The graph of the power equation for the wood pigeon, with flight velocities for minimum power (V_{mp}) and maximum range (V_{mr}) is shown in (b). Power requirements for hovering and low flight velocities (broken curve) are estimated from sources other than the power equation.

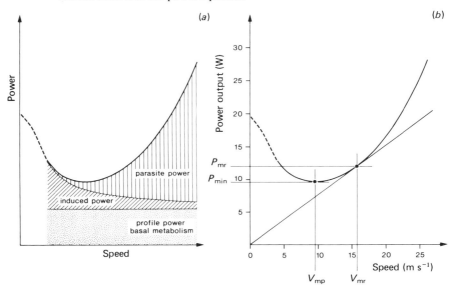

on radar tracking of only five flocks. These swan flocks were migrating low over the Baltic Sea at the end of April towards their northeasterly breeding quarters. On repeated occasions flocks of swans were observed swimming over deep water in their migratory direction, and one flock of 22 whoopers tracked by radar landed on the water surface and the birds swam purposefully for almost two hours before they set off flying again, and radar tracking was resumed.

This may suggest that swans have difficulty in producing the power required for continuous flight, and that they are forced to fly with minimum power output. Power required for flying in birds with different body masses (M) approximately increases as $M^{7/6}$, while the capacity of power output depends on the mass of flight muscles increasing approximately in proportion to M. This implies a maximum mass in birds above which flight is not possible, and calculations suggest that this size approximates to that of the largest flying birds: about or slightly above 10 kg, for example, for swans, bustards, albatrosses and condors. Hence, it may well be difficult for the whooper swans, presumably loaded with some fat reserves for their migratory journey, to muster up enough power to fly at V_{mp} and it may be impossible for them to fly continuously at V_{mr}.

The birds' adjustment of their flight speed is nicely illustrated in the swift (Bruderer & Weitnauer, 1972). While the mean air speed is 11 m s^{-1} in migrating swifts, it is only 6 m s^{-1} in breeding swifts spending the night aloft. Indeed,

Table 1. *Observed air speeds in migrating birds, studied by radar, in comparison to speeds for minimum power* (V_{mp}) *and maximum range* (V_{mr}) *as predicted theoretically from the power equation*

	Mass (kg)	Wing span (m)	V_{mp} (m s^{-1})	V_{mr} (m s^{-1})	Observed mean air speed (m s^{-1})
Chaffinch *Fringilla coelebs*	0.020	0.30	5.1	9.5	11
Swift *Apus apus*	0.043	0.45	5.4	9.4	11
Redwing thrush *Turdus iliacus*	0.065	0.40	6.6	11.5	13
Wood pigeon *Columba palumbus*	0.50	0.75	9.5	15.7	17
Oystercatcher *Haematopus ostralegus*	0.55	0.80	9.5	15.7	14
Eider duck *Somateria mollissima*	2.0	1.1	12.5	20.4	21
Bean goose *Anser fabalis*	3.5	1.6	12.5	20.3	20
Crane *Grus grus*	5.5	2.4	11.8	19.4	19
Whooper swan *Cygnus cygnus*	10	2.4	14.4	23.5	17

Observed speeds refer to horizontal flapping flight. Hence, for the crane, which migrates by soaring over land, the observed speed refers to birds migrating across the Baltic Sea. Data for the chaffinch and swift are from Bruderer (1971) and Bruderer & Weitnauer (1972), respectively, and the rest from Alerstam (1976, and unpublished).

these values fall close to the predicted speeds with energy consumption minimal per unit distance (V_{mr}) and per unit time (V_{mp}), respectively (Table 1). Furthermore, white-throated sparrows *Zonotrichia albicollis* released experimentally and tracked by radar, under clear night skies fly off in their normal migratory direction at a mean air speed of 9.2 m s^{-1}, similar to V_{mr}, but when released under overcast conditions, they fly on irregular tracks, with obvious difficulties of orientation, at 6.6 m s^{-1} mean air speed, close to V_{mp} (Emlen & Demong, 1978).

There are some predictions about the optimal adjustment of air speeds in migrating birds that remain to be critically tested (Pennycuick, 1978). One is that V_{mr} increases with increasing flight altitude, another that V_{mr} declines with increasing tailwinds and increases with headwinds. The ground speed (G) of a migrating bird is the vector sum of the bird's air speed and wind speed. Hence, the power equation may be used to calculate V_{mr} under different wind conditions by solving $dP/dV = P/G$. Both radar and optical measurements strongly indicate that birds indeed do fly relatively faster in headwinds than in tailwinds (Bellrose, 1967; Bruderer, 1971; Tucker & Schmidt-Koenig, 1971; Emlen, 1974; Schnell & Hellack, 1979).

Soaring migration

As an alternative to continuous flapping flight many birds adopt a strategy of soaring migration. They exploit lift in winds deflected upwards at hills and escarpments, so-called slope soaring. Even more important is thermal soaring, where birds take advantage of warm, rising air in an unstable air mass. The irregular heating of the ground by the sun leads to a narrow turbulent layer close to the ground with downdraughts and incipient thermals. Most of these thermals lose their identities in the general convective confusion in this layer, but a few develop into mature thermals, large air bubbles of a vortex-ring type with cores which may be one or two kilometres across at high altitude. When thermals reach their condensation level they give rise to cumulus clouds. Just like glider pilots, birds climb in thermals by circling at their centre, where the lift is best; on reaching the altitude at which the thermals weaken the birds glide off in their migratory direction, later to exploit a new thermal (Fig. 2). There is often considerable lift extending inside cumulus clouds but birds seem mostly to break off their climb at the cloud base. However, more detailed studies to determine whether birds sometimes use the lift right inside clouds would be very interesting.

A gliding bird sacrifices potential energy to support its weight and overcome the drag of body and wings. The rate of potential energy loss is directly proportional to the sinking speed, V_z, while the power component required for lift decreases proportionally to V^{-1} (V is the forward gliding speed) and that for thrust increases with V^3. Hence, by analogy with the derivation of the power equation, sinking speed in gliding flight is related to gliding speed according to the equation

$$V_z = \alpha V^{-1} + \beta V^3$$

Fig. 2. Thermals grow in a heated ground layer, from which they soon break away to form isolated bubbles of rising air with vortex ring motion (*a*). These thermals are exploited by soaring birds, which circle in them to gain height, and glide away to another thermal in cross-country soaring migration (*b*).

(*a*)

(*b*)

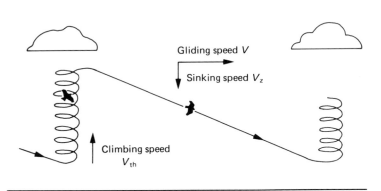

where α and β are negative constants related to the bird's mass, wing area and wing span, and to air density and coefficient of drag (Pennycuick, 1972a, 1975). The graph of this relationship is called the glide polar (Fig. 3). The glide polar shows three characteristic gliding speeds. One is the stalling speed, or minimum gliding speed, V_{min}, which generally is smallest for birds with low wing loading (body mass divided by wing area). V_{ms} defines the gliding speed with minimum sink, while V_{bg} gives the best glide ratio, i.e. the ratio of forward gliding speed to sinking speed, which of course is equivalent to the ratio of horizontal gliding distance in still air to height loss. Good gliding performance in birds, both in terms of a low sinking speed and a high glide ratio, is closely connected with a high aspect ratio (ratio of wing span to chord).

Albatrosses, with their extremely long and slender wings, have the highest glide ratios among birds, about 24:1, while raptors, which frequently soar, have best glide ratios ranging from about 10:1 to 15:1. These figures are poor in comparison with modern gliders which achieve glide ratios close to 50:1. However, it is not only gliding performance that counts when it comes to soaring migration. Climbing performance is also important and is best in birds with a low wing loading, since they can bank and fly in circles of small radius without increasing sinking speed too much. By soaring in tight circles they can effectively exploit the core of best lift in the thermals and, since a low wing loading also gives a low minimum sinking speed (see Fig. 3), these birds are particularly fitted to use weak and narrow thermals. However, under good soaring conditions with strong and wide thermals a low wing loading is a disadvantage because the cross-country speed will be low. Under these conditions, the birds with the highest wing loadings will go the fastest.

The overall cross-country speed (V_{cc}) in soaring migration can be calculated from

$$V_{cc} = VV_{th} (V_{th} - V_z)^{-1}$$

where V is forward gliding speed, V_z sinking speed (negative) and V_{th} is the climbing speed that will be achieved in the thermals (Fig. 2). By rearranging the expression above, one finds that there is a useful graphical interpretation of this relationship (Fig. 3(b)). Hence, one can easily see that the interthermal gliding speed for obtaining maximum cross-country speed is always higher than V_{bg}, and the optimal gliding speed (V_{opt}) increases with increasing climbing speed in the thermals. If soaring birds are anxious to maximize their cross-country speed one would therefore expect them to keep adjusting their interthermal gliding speed in relation to the lift experienced in the preceding thermals. However, if strong thermals are widely scattered, it will be better always to glide at V_{bg}, thereby maximizing the distance between thermals. These are the extreme strategies for pure soaring migration, since a bird seeking a compromise between high cross-

Fig. 3. (a) The glide polar is the graph of sinking versus forward speed in gliding flight. Minimum gliding speed (V_{min}), and gliding speed with minimum sink (V_{ms}) and best glide ratio (V_{bg}) are characteristics of the polar. (b) Glide polar estimated in the crane. The average cross-country speed (V_{cc}) is found as the intercept on the abscissa by drawing a straight line from the achieved rate of climb in thermals on the ordinate (V_{th}) to the relevant gliding speed on the glide polar. Hence, the optimal gliding speed V_{opt} for maximizing overall cross-country speed is found by drawing a tangent to the polar from V_{th}, as shown in the figure. V_z, sinking speed. During a study of spring migration of cranes in South Sweden, mean climbing rate in thermals was about 1.5 m s^{-1}. As seen from the figure, this gives a maximum cross-country speed at about 10 m s^{-1} in pure soaring migration. However, the cranes usually travelled faster, 11–14 m s^{-1} because they used varying amounts of flapping flight on the interthermal glides.

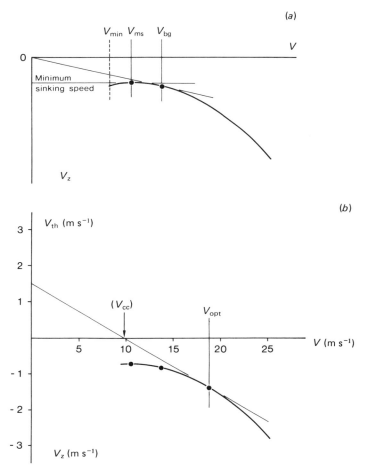

country speed and reliability of finding thermals would glide at a speed between V_{bg} and V_{opt} (Welch & Irving, 1977; Pennycuick, 1978).

White storks travel by cross-country soaring between Europe and their winter haunts in Africa, an average distance of about 7000 km. Their soaring behaviour during migration in East Africa was observed from a motor-glider by Pennycuick (1972a). The storks travel in large flocks, typically with 200–500 individuals. They do not fly in formation during circling, and a circling column usually contains individuals flying in both directions. Between thermals the individuals space themselves out and glide along parallel headings in loose formation, with the flock extending 200–300 m laterally. When some individuals come through a thermal the whole flock quickly converges on the birds which have started to climb. During this process the storks seem not to pay any attention to visible signs of good thermals, like growing cumulus clouds, but to rely on the efficiency of locating lift somewhere in the laterally extended flock. Stork migration in East Africa takes place mainly in the height band 1000–2500 m above the ground, corresponding to 2500–4000 m above sea level. The storks glide at up to 18 m s^{-1} between thermals, and their average cross-country speed is 12 m s^{-1}, which is consistent with predictions from the stork's glide polar under favourable thermal conditions.

Cranes frequently adopt soaring migration on their journeys between winter grounds in the southwestern Mediterranean area or along the Nile and breeding sites in the northern taiga swamps, and their behaviour has been studied both by radar and by following them in a light aircraft (Alerstam, 1975a; Pennycuick, Alerstam & Larsson, 1979). Fig. 4 shows an aircraft track of a migrating crane

Fig. 4. Record of height at the start and finish of each climb in a thermal versus time for a flock of migrating cranes tracked by a light aircraft (from Pennycuick et al., 1979).

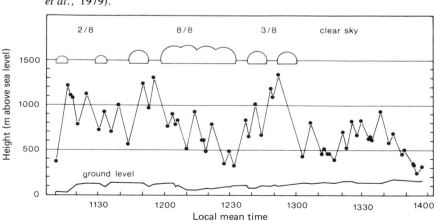

flock, and referring to this figure, I cite from Pennycuick *et al.* (1979): 'Soaring conditions were quite varied on this flight. Initially thermals were good, under a sunny sky with small cumulus clouds, and the cranes achieved a series of good climbs in the height range 500–1300 m a.s.l. Then they entered an area where the cumulus clouds had coalesced and cut off the sun from the ground ("over-convection"). Thermals were weaker here, and the cranes progressively lost height to about 300 m, but made their way back to 1300 m soon after emerging from the overcast area. Later the cumulus clouds ceased altogether as the air became stable, and the cranes again progressively lost height in the weak, inconsistent thermals. When we left them, they were within 150 m of the ground and obliged to resort to flapping flight'.

The migrating cranes behave differently from storks in several respects. In a thermal all cranes in a flock, usually comprising 10–100 individuals, circle in the same direction in a regular box formation, and when flying straight they maintain a V- or J-formation. When convection is strong, cranes glide between thermals, flapping their wings only occasionally to maintain position. However, cranes are certainly not totally dependent on soaring, but when at low altitude and/or when thermals are poor, gliding is interspersed with flapping flight, and the birds may even cover considerable distances in continuous flapping flight. This happens over the open sea, where thermals are absent, when cranes travel across the Mediterranean (which storks circumfly over land) and the Baltic Sea. During continuous flapping flight the cranes' speed is close to V_{mr}, as predicted from the power equation (Table 1). Hence for cranes, in clear contrast to white storks, powered flight may be the primary mode of migration, and available thermals are used as a supplementary source of energy to reduce the cost of transport. Under favourable soaring conditions the cranes achieve cross-country air speeds between 10 and 13 m s^{-1} as predicted from the glide polar (Fig. 3). However, with poor thermals this speed is not reduced, but is actually slightly higher, because of the significant fraction of interspersed flapping flight under such conditions.

Migration of soaring birds attracts much attention among bird-watchers because of spectacular concentrations of these migrants at passages with favourable soaring conditions. The migrants make long detours to avoid having to use flapping flight over the sea, and the sites most famous for soaring bird migration in Europe, Falsterbo, Gibraltar and the Bosphorus, are situated at minimal sea-crossings. Annual numbers of soaring autumn migrants counted at these sites are given in Table 2. Depending on the relative degree of concentration to these sites, I have divided the species into three categories. Storks, eagles and buzzards are most prone to make long detours to minimize sea-crossings, while this habit is intermediate in sparrowhawks and harriers, and not very strong in the osprey and the falcons. The soaring migrants passing Falsterbo are recruited from breeding populations in Sweden and, for a few species, also Finland, and sizes of these populations are fairly well known. Up to 30% of the common buzzard population from Sweden may be counted by the Falsterbo observatory (Rudebeck, 1950). On the basis of the ratio between numbers of migrants observed at Falsterbo and size of the recruiting population, a relative index of concentration can be estimated at 100 in the buzzards, about 20 in the sparrow-hawk and harriers, and five in the osprey and falcons.

Assuming that the power required to hold a bird's wings in gliding position is approximately equal to the basal metabolism, on the basis of the glide polar and power equation one can compare energy consumption per unit distance covered in cross-country soaring (with average to strong thermals) with that in continuous flapping flight (Pennycuick, 1978). It follows that the energetic advantage of soaring migration in relation to powered flight is positively correlated with the bird's mass and inversely correlated with the wing span. Hence, in buzzards the energetic cost per unit distance covered by flapping flight is about three times higher than cost by soaring migration, and this factor is even higher in eagles and storks. However, in sparrowhawks, harriers and falcons the energetic advantage of soaring is distinctly less, the cost of transport being about half of that in flapping flight. Hence, from an energy point of view it will be favourable for buzzards to avoid extensive sea crossings by making a detour over land up to three times longer than the direct oversea route. In the white stork such a detour may be energetically favourable even if it is up to six times longer than the direct route. In contrast, sparrowhawks, harriers and falcons should resist detours that are twice, or more, as long as the oversea route. This helps to explain why the latter birds do not concentrate so strongly at narrow sea crossings. However, it neither explains why sparrowhawks and harriers undertake sea crossings less readily than falcons, nor why ospreys (and cranes) make detours via the shortest sea crossings so rarely. The latter fact is indeed astonishing, since the ospreys and cranes should save 70–90% of the costs of transport by adopting cross-country soaring instead of powered flight.

There are a number of disadvantages in soaring migration as compared to powered flight. Ground speed is less in soaring migration, and this strategy brings restrictions on the choice of weather, time of day (soaring is usually restricted to the hours around noon, when thermals are fully developed), and routes for the migratory flights. Hence, even for large birds with good climbing and gliding performance, like the osprey and crane, these disadvantages may balance the energetic advantage of soaring migration to the extent that sea cross-ings during migration are avoided only to a small degree or not at all.

During autumn migration, Scandinavian ospreys spread across Europe and travel on an extremely broad front to their African winter quarters. Ringing recoveries during autumn migration range from Britain and Portugal in the west

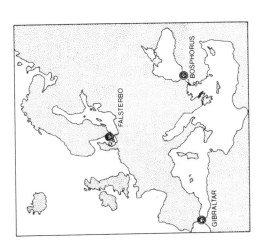

Table 2. *Annual numbers of soaring birds migrating past Falsterbo, Gibraltar and the Bosphorus in the autumn.* The relative degree of concentration of migrants to these narrow sea passages (cf. map) is high in category A species, intermediate in category B and small in category C. Counts were performed at one observation site at Falsterbo as well as the Bosphorus, while several observation sites were manned in the Gibraltar region. Falsterbo figures are rounded mean annual totals during five years in the seventies (Roos, 1978a) and maximum annual totals in 18 autumns 1942–1977. Gibraltar figures are means of counts during two autumns, 1972 and 1974 (Bernis, 1975). For the Bosphorus, total numbers during autumn 1966 (Porter & Willis, 1968) are given together with maximum numbers recorded in 8 autumns 1966–1973 according to the Bird Reports of the Ornithological Society of Turkey. Numbers less than ten are omitted. Numbers of sparrowhawks and kestrels at the Bosphorus are rough estimates, because of identification difficulties in relation to Levant sparrowhawks and lesser kestrels, respectively.

Species	Falsterbo[a]	Gibraltar	Bosphorus[a]
A. White stork *Ciconia ciconia*	–	15 000	207 000 (340 000)
Black stork *Ciconia nigra*	–	200	6 000 (8 000)
Egyptian vulture *Neophron percnopterus*	–	3 000	400 (600)
Booted eagle *Hieraaetus pennatus*	–	10 000	300 (500)
Short-toed eagle *Circaetus gallicus*	–	7 000	1 300 (2 300)
Lesser spotted eagle *Aquila pomarina*	–	–	4 000 (19 000)
Black kite *Milvus migrans*	–	30 000	2 200 (2 700)
Honey buzzard *Pernis apivorus*	7 000 (22 000)	90 000	9 000 (26 000)
Common buzzard *Buteo buteo*	11 000 (37 000)	2 800	13 000 (33 000)
Rough-legged buzzard *Buteo lagopus*	500 (1 200)	–	–
B. Sparrowhawk *Accipiter nisus*	6 000 (11 000)	700	500
Levant sparrowhawk *Accipiter brevipes*	–	–	5 000 (6 000)
Hen harrier *Circus cyaneus*	130 (220)	50	–
Marsh harrier *Circus aeruginosus*	80 (110)	250	(10)
Montagu's harrier *Circus pygargus*	–	1 200	(10)
C. Osprey *Pandion haliaetus*	90 (110)	50	(10)
Kestrel *Falco tinnunculus*	240 (420)	1 000	50
Hobby *Falco subbuteo*	20 (40)	200	80 (100)
Merlin *Falco columbarius*	90 (220)	50	–

[a]*Maximum numbers in parentheses*

to the Caspian Sea in the east (Österlöf, 1977). The extensive dispersal during the southward migration is possibly due to intraspecific competition at bays, lakes and rivers where the migrating ospreys rest and fish. These selection pressures of course also affect the osprey's propensity for making detours by soaring migration, thereby causing them to become concentrated in large numbers in restricted regions.

Soaring migration is used not only by landbirds but also by ocean ones like petrels, fulmars and shearwaters. These birds soar in slope lift over the sea waves, and to some extent they also extract energy from the gradient of horizontal wind speed over the sea surface, so-called dynamic soaring (Pennycuick, 1972b). Albatrosses are the real experts in the latter technique. Due to friction, horizontal wind speed is lower close to the sea surface than higher up. By gliding into the wind when climbing, the albatrosses extract lift from this wind gradient, and by turning downwind on the downward glide they gain extra kinetic energy from the wind gradient during this phase.

Fuel economy

Birds store fat for use as fuel during their migratory flights. Depot fat is the most economical type of fuel in terms of a high oxidizing energy per unit mass. The fuel ratio (F) is the fraction of take-off mass of a migrating bird that consists of fuel, and it may be as high as 50% in birds performing long non-stop flights. Taking into account the energy content of the fat load, one can estimate the maximum flight range in still air for birds with different fuel ratios, as seen in Fig. 5 (cf. Pennycuick, 1975). Large birds do not store fat to extremely high fuel ratios, because, with too much extra weight, they cannot attain the muscle power equivalent to P_{mr}.

Flight range is greatly affected by winds, and therefore it is especially useful to estimate the total time a bird can spend aloft flying at V_{mr} on a given load of fuel. On the basis of the power equation, the expected relative weight loss due to consumption of fat is about 0.7% body mass per flying hour in small birds like warblers, finches and thrushes, and up to 1–1.5% in large birds like ducks, geese and swans. The rate of weight loss during flight will be less in birds with a wide wing span than in birds with short wings.

Various field estimates of the rate of weight loss in migrating birds are presented in Table 3. Excluding the value for Manx shearwater, which uses soaring migration (see below), the estimates fall between 0.6 and 2% loss of body mass h^{-1}. Some values agree closely with the theoretical predictions, while others are surprisingly high. However, all estimates depend on more-or-less uncertain assumptions and in view of the widely different methods, the variation between the estimates is not unduly high (Nisbet, 1963a). It may be that interspecific differences in effectively storing and using fat are more important than expected from theory, or that a high rate of weight loss in some birds is due to excessive loss of water under certain conditions. Very little is known about water economy and the risk of dehydration in migrating birds (Hart & Berger, 1972).

The time a bird with a certain fuel ratio can fly until all fuel is used up is given by

$$t = k^{-1}\ln(1-F)$$

where k is the relative weight change due to fuel consumption per flying hour. Hence, with $F = 0.5$ and $k = -0.007$, maximal flying time is about 100 h. These conditions are relevant for e.g. the blackpoll warbler travelling from northeastern North America directly across the Atlantic Ocean to winter quarters in northern South America (Nisbet, 1970). On the basis of radar observations over the Atlan-

Fig. 5. Maximum flight ranges in still air for birds with different fuel ratios. The straight lines of flight ranges in birds of different sizes are rough approximations only, since the birds' dimensions are important for the energy consumption in flight. From upper to lowest lines, $F = 0.5$, 0.25, and 0.1, respectively. By way of example, birds with a relatively wide wing span attain longer flight ranges than birds with short wings. Calculations are based on Pennycuick (1975).

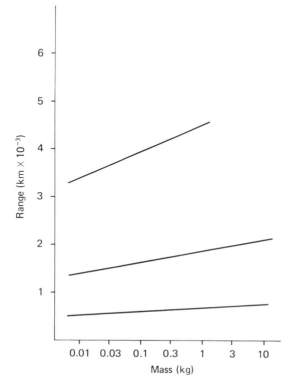

Table 3. *Field-estimates of the rate of weight-loss in migrating birds*

Species	Weight-loss (% body mass h^{-1})	Basis of field estimate
1. Blackpoll warbler *Dendroica striata*	0.6	Estimated arrival and departure weight for non-stop flight: from New England to Bermuda
2. Robin *Erithacus rubecula*	0.9	from Norway to Scotland
3. Goldcrest *Regulus regulus*	0.7	from Norway to Scotland
4. Wheatear *Oenanthe oenanthe*	0.8	from Greenland to Scotland
	1.3	
5. Knot *Calidris canutus*	1.0	from England to Iceland and from Iceland to Ellesmere Island
6. Manx shearwater *Puffinus puffinus*	0.3	from Wales to Brazil
7. Song sparrow *Melospiza melodia*	0.7	Weight before and after a night of migration, Massachusetts
8. Thrushes *Hylocichla* spp.	1.8	Weight at different times in the night: Migrants killed at TV-tower, Illinois
9. Tennessee warbler *Vermivora peregrina*	1.8	Migrants killed at TV-tower, Wisconsin
10. Veery *Hylocichla fuscescens*	1.3	
11. Ovenbird *Seiurus aurocapillus*	1.0	Migrants killed/trapped at lighthouse, Ontario
12. Chaffinch *Fringilla coelebs*	2.9 (2.0)	Weights at trapping sites 50 km apart along the migratory pathway, East Baltic

Note: The original estimate for the chaffinch was based on an assumed air speed at 14.5 m s^{-1}. A mean air speed at 10 m s^{-1} seems to be more realistic in this species (Table 1) and gives an estimated rate of weight loss shown in brackets. Studies 1–4 and 7–8 are reviewed by Nisbet (1963a), 5 refers to Glutz von Blotzheim, Bauer & Bezzel (1975) and Morrison (1977), 6 to Perrins *et al.* (1973), 9 to Raveling & Le Febre (1967), 10 and 11 to Hussel (1969) and 12 to Dolnik & Gavrilov (1973).

tic Ocean, a normal duration of 70 to 100 hours is estimated for birds undertaking this formidable non-stop flight (Larkin, Griffin, Torre-Bueno & Teal, 1979). The above calculation is relevant for a bird flying at V_{mr}; if the bird slows down to V_{mp} it may stay in the air for about 130 hours under the same conditions. Actually, a lot of radar echoes recorded at relatively low altitude over the western Atlantic ocean, which probably were from migrating birds, showed air speeds at or below 5 m s^{-1}, i.e. close to V_{mp} in small birds (Williams, Williams, Ireland & Teal, 1977; Larkin et al., 1979). Perhaps migrants over the ocean were slowing down to save energy, while waiting for more favourable weather and better conditions for orientation?

One of the most impressive feats in animal behaviour is the migration of birds across thousands of kilometres of ecological barriers, like oceans, glaciers and deserts. In the light of today's knowledge of such enormous non-stop journeys, the classical case of migration across the North Sea is no longer so impressive. The distance across the North Sea from Scandinavia to Britain is 500–800 km and the normal flight time is between 10 and 20 hours for small birds. The North Sea is regularly crossed by a multitude of bird species, passerines, shore birds and waterfowl, and the radar studies by David Lack (1959–1963) have demonstrated the very vivid traffic of migrating birds over this region.

Another classical example of a long sea crossing by birds is the migration across the Gulf of Mexico, a distance of about 1000 km. For small birds, this flight normally lasts about 20 h, in poor weather considerably longer (Lowery, 1945; Gauthreaux, 1971). The trans-Gulf migrants, including the ruby-throated hummingbird *Archilochus colubris,* attain fuel ratios up to 45% (Odum, Connell & Stoddard, 1961), making it possible for the birds to fly a distance, without refuelling, that is at least twice as long as the actual Gulf-crossing.

In Fig. 6 are shown some of the longest non-stop flights by birds in the North Atlantic area. By way of example, snow buntings *Plectrophenax nivalis,* breeding on East Greenland, cross the Greenland Sea on their way to winter quarters in Russia, while birds of West Greenland populations fly across Davis Strait to North America. Brent geese *Branta bernicla,* white-fronted geese *Anser albifrons,* turnstones *Arenaria interpres,* knots *Calidris canutus* and wheatears *Oenanthe oenanthe* breeding in Greenland and parts of the Canadian archipelago fly to and from western Europe, either via Cape Farwell or across the ice-cap of Greenland, sometimes but not always using Iceland as a stop-over site (Salomonsen, 1967). In the autumn, knots travel directly from northeast Greenland to the North Sea countries (Morrison, 1977). In spring, knots in England store a fuel ratio at over 30% for the flight to Iceland. On Iceland the knots refuel, and birds heading for Ellesmere Island attain fuel ratios at 40%. Practically all fat has been consumed when they reach their destination. The flight of the above species across the Greenland ice, extending 2000–3000 m above sea level with temperature at −10°C, surely is a formidable achievement.

Ringing recoveries indicate that during the autumn Greenland wheatears often travel directly to France or the Iberian peninsula, a distance of approximately 3500 km. The fuel ratio is about 50%. The birds probably use northwesterly tailwinds on the rear side of a cyclone for this journey. The wheatears' ultimate destination is their winter quarters in tropical Africa (Salomonsen, 1967).

A regular migration route from North to South America directly over the western Atlantic Ocean, a distance between 3000 and 4000 km, has recently been documented by observations from a network of radar stations: marine, surveillance and tracking radars (Williams *et al.*, 1977). This route is used not only by the blackpoll warbler as mentioned above, but also by several other species of songbird, and quite a number of shorebird species (McNeil & Burton, 1977). A great many shorebirds depart southeastwards in the autumn from staging areas at Nova Scotia. Furthermore, it regularly happens in this region that shorebird flocks arriving from far inland continue non-stop over the ocean (Richardson, 1979). Fuel ratios reach 50% in the waders resting at Nova Scotia.

Just to mention some examples of long non-stop flights over oceans other than the Atlantic, brent geese fly almost 4000 km across the Pacific Ocean from the Alaska Peninsula to the South Californian coast (Ogilvie, 1978); two species of New Zealand cuckoos fly about 3500 km to the Solomon and Samoa Islands, respectively (Lack, 1959); passerines, bee-eaters and Amur falcons traverse

Fig. 6. Examples in the North Atlantic region of routes that are regularly used by migrating birds in non-stop flights.

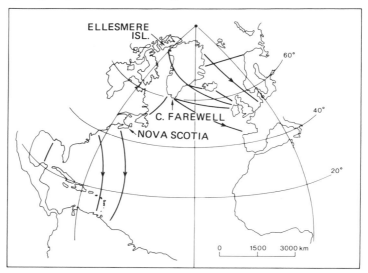

almost 3000 km of the Indian Ocean between India and East Africa (Moreau, 1972).

Many migrants fly between Europe and Africa across the Mediterranean Sea and the Sahara desert in one single non-stop flight, lasting at least 40 h in the autumn, when winds generally are favourable, and 60 h in spring when they are less favourable (Moreau, 1972). Fuel ratios at 35–40% are normal in passerines undertaking this journey (Ward, 1963; Fry, Ash & Ferguson–Lees, 1970). There is also circumstantial evidence that many songbirds fly non-stop from Kazakhstan in the Soviet Union, across the Trans-Caspian and Arabian deserts, to Ethiopia and Sudan, a distance of almost 4000 km (Moreau, 1972).

Bird migration in hospitable regions, with many suitable foraging and resting sites, normally does not proceed by such magnificent non-stop flights. Rather, the birds travel by numerous short flights, each of about three to ten hours duration. Fuel ratios are often about 15–20%. Nocturnal migrants depart soon after sunset, and diurnal migrants set off around sunrise to migrate during the morning hours. Many species depart either by day or by night, whenever weather becomes favourable. Flying with a small load of fat is advantageous, since the costs for transportation of extra fat will be low.

However, there are also drawbacks involved in migrating by numerous short flights and with low fuel ratios. The birds will have to refuel often, and many studies have demonstrated that migrants do not start gaining weight until some days, typically two or three, after their arrival at a new resting site (cf. Mascher, 1966). Northern waterthrushes Seiurus noveboracensis defend territories at stopover sites. Territory holders show a weight gain, while newly arrived migrants, which have not yet attained a territory, tend to lose weight. It normally takes two or three days before a newcomer attains a territory (Rappole & Warner, 1976). Hence, migrants lose time at the resting sites, and they may also expose themselves to various risks in the new environments.

There are species that undertake long non-stop flights also over seemingly favourable terrain. In Britain, sedge warblers Acrocephalus schoenobaenus put on fat up to a fuel ratio above 40% prior to the autumn migration, indicating that these birds fly directly to West Africa without refuelling (Gladwin, 1963). Geese frequently travel long distances non-stop over land, like the lesser snow geese Anser caerulescens in North America, flying from James Bay to the coast of the Gulf of Mexico. This distance is about 2700 km and is covered by the geese in less than 60 h (Ogilvie, 1978). Waders also travel by extremely long flights, passing many suitable potential resting places. Dunlins Calidris alpina in England attain a fuel ratio of 40% prior to spring migration towards the northeast across the North and Baltic Sea regions (Pienkowski, Lloyd & Minton, 1979). Various wader species wintering in South Africa put on similar amounts of fat before departing northwards. Hence they probably cover the total distance to their high arctic Eurasian breeding grounds, about 13 000 km, in four or so long flights (Summers & Waltner, 1979).

Little is known about the fuel economy in birds using soaring migration. Adult Manx shearwaters desert their fat chicks at the breeding islands. Average chicks have a fuel ratio close to 40% and it is probable that many young birds cover the autumn flight from Britain to Brazil, almost 10 000 km, without refuelling. One young bird was recovered on the coast of Brazil a fortnight after being ringed in Britain (Perrins, Harris & Britton, 1973). The shearwaters' method of flight over the sea, a combination of slope and dynamic soaring, requires much less energy than normal flapping flight. The relative weight loss per flying hour in shearwaters is only about one-third or one-quarter of that in birds using flapping flight (Table 3).

Honey buzzards *Pernis apivorus* migrate by cross-country soaring, exploiting thermals over land. During summer, a honey buzzard weighs on average 625 g, but prior to autumn migration the birds put on fat to attain a weight of about 900 g (Glutz von Blotzheim, Bauer & Bezzel, 1971). Assuming that this increment in mass corresponds to fuel used up during the migratory journey, and that the buzzards travel with a mean ground speed of about 13 m s^{-1} over a distance of at least 7000 km from Europe to tropical Africa without refuelling, the weight loss will be about 0.3% body mass per flying hour. This is about one-quarter of the energy consumption required for flapping flight.

In contrast to the honey buzzards, common buzzards *Buteo buteo vulpinus*, covering the 10 000 km journey between East Europe/West Siberia and South Africa by soaring migration, do not store large reserves of fuel prior to their migration. Between their arrival at South African winter quarters in October/November and their departure northwards in February, the common buzzards gain less than 10% in weight (Broekhuysen & Siegfried, 1971). Presumably, the migrating common buzzards spend morning and evening hours, before and after the daily period available for thermal soaring, hunting for small rodents. The total duration of the common buzzards' journey is about two months; during March and October they migrate through Africa and in April and September they pass through the region of the Black and Caspian Seas.

Migratory route

The shortest route between two points on the earth's surface runs along the great circle. In order to follow a great circle, a bird must continuously change its flight course. A more convenient route for navigation is the path of constant compass heading between two points: the rhumbline. Hence, the rhumbline intersects all longitudes at the same angle and is represented as a straight line on a Mercator map projection.

For birds travelling long distances in east-west direction at mid and high latitudes the great circle route brings a considerable reduction in distance, and consequently in flying time and energy costs, as compared to the rhumbline path. There are a few cases that convincingly show that the birds' migratory route under such conditions falls close to the great circle.

Large numbers of ruffs *Philomachus pugnax* pass through northwestern Europe on their autumn migration. Their principal destination is the western part of the Sahel, in the inundation areas at the rivers Senegal and Niger and at Lake Chad. Ringing recoveries show that some of these birds originate from breeding areas in East Siberia, around and even east of the river Lena. The great circle and rhumbline between Lena (70° N, 125° E) and Senegal (15° N, 15° W) are shown in Fig. 7a. The autumn passage through northwest Europe of ruffs from so far to the east indicates that the birds follow close to the great circle, which runs exactly across this region. Birds following the great circle should depart from Lena towards approximately northwest (322°) and gradually change their flight direction to arrive at Senegal on a final course between southsouthwest and south (193°). The great circle distance is 10 060 km, while the rhumbline route (towards 239°) is 18% longer. It is not known whether the ruffs fly out across the Arctic Ocean to follow the great circle in detail – they may choose the slightly longer route along the arctic coast to have access to a great number of suitable resting sites.

On spring migration the ruffs probably travel along a route between the great circle and the rhumbline – the main stream of migrants now flows across Italy and Central Europe. The birds are probably prevented from flying the shorter route over northerly areas in spring, because there are no resting sites at this time of the year, when the tundra is still frozen.

During autumn another arctic wader species, the curlew sandpiper *Calidris ferruginea* with its most important breeding area on the Taimyr Peninsula, regularly passes the coasts of West and Northwest Europe, close along the great circle route to important winter quarters at the Mauritanian coast. The great circle distance of this journey is 8540 km, while the rhumbline is 12% longer (Fig. 7(b). Significant numbers of curlew sandpipers also winter close to Cape Town, in the southernmost part of South Africa. Recoveries of birds ringed in this region are reported around the Caspian and Black Seas, clearly west of the rhumbline towards Taimyr but close to the great circle. However, the rhumbline distance is only between two and three per cent longer than the great circle distance (13 460 km), and it may well be the availability of favourable resting sites at the Caspian and Black Seas, rather than the reduction in flight distance, that is the most important selecting factor behind this migratory route.

Winds have a profound effect on the ground speed, and energy costs, of the birds' flights. The distribution on earth of solar heating in combination with the earth's rotation, bring about a characteristic pattern of wind and pressure zones

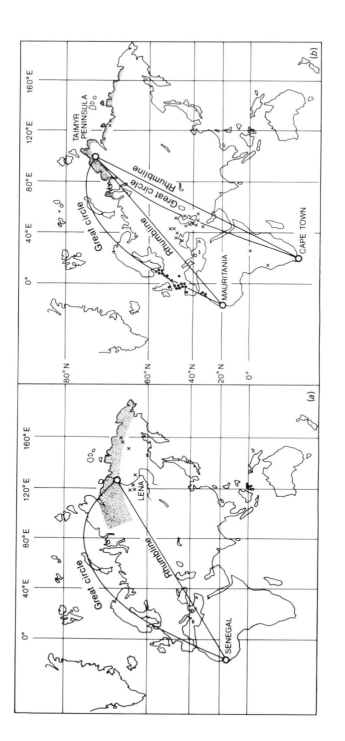

on earth. At middle latitudes (about 30°–60°) west winds, frequently associated with passages of cyclones, dominate on both hemispheres, while easterly trade winds prevail closer to the equator, and polar easterlies blow in the extreme north and south (Fig. 8(a)).

An association between migratory routes and following winds is particularly evident in many sea birds. This is exemplified by the great shearwater *Puffinus gravis* in Fig. 8. This species breeds, in millions of pairs, on two islands in the Tristan da Cunha group, and on Gough Island, in the South Atlantic Ocean. After breeding the birds depart northwestwards with following trade winds past the horn of Brazil and across the Sargasso Sea into the west wind belt off New-foundland. The great shearwaters 'winter' in the North Atlantic west wind zone. Their main distribution during the northern summer is gradually shifted east-wards (with the winds) from the Banks off Newfoundland via Greenland offshore waters, where the birds undergo a rapid moult, to Rockall Bank and West Euro-pean waters where they appear between August and October. Fish constitute important food in this species, and the shift of shearwaters from Newfoundland to Greenland waters coincides with a shift in distribution of spawning capelin.

A clockwise route of migration in the northern hemisphere, maximizing the benefit of following winds, is also followed by the sooty shearwater *Puffinus griseus* and Wilson's storm-petrel *Oceanites oceanicus* in the Atlantic Ocean, and the short-tailed shearwater *Puffinus tenuirostris* (Dorst, 1962) and south polar skua *Catharacta maccormicki* in the Pacific Ocean (Devillers, 1977), for example.

Manx shearwaters *Puffinus puffinus* breeding in West Europe migrate south-west through the trade wind belts to waters off the east coast of South America, probably continuing by a counterclockwise movement in the South Atlantic, thereby making maximal use of tailwinds in the southern hemisphere (Cramp & Simmons, 1977). Wind is a key factor also for the movements of albatrosses, circling the south pole with tailwinds within the west-wind belt over the Southern Ocean (Dorst, 1962).

Fig. 7. Great circle and rhumbline routes for (a) ruffs *Philomachus pugnax* migrating between the rivers Lena and Senegal and (b) curlew sandpipers *Calidris ferruginea* migrating between Taimyr Peninsula and Mauritania or Cape Town in Africa. In (a) the stippled area indicates the breeding range in East Siberia and crosses mark recoveries of birds ringed during migration through Finland, Sweden, Denmark, Holland, Germany and England. In (b) the stippled area indicates the breeding range, circles mark autumn recoveries of birds ringed during migration through Sweden, and crosses mark spring and autumn recoveries of birds ringed during winter in South Africa. The routes are plotted on a Mercator map projection – note that areas close to poles are disproportion-ately enlarged in this projection. Ringing and distribution data are from Glutz von Blotzheim *et al.* (1975) and Elliott *et al.* (1976).

However, sea-bird migration is certainly not always directed downwind –
other selecting factors may override the positive effect of following winds. For
example, large numbers of fulmars *Fulmarus glacialis*, kittiwakes *Rissa tridac-
tyla* and little gulls *Larus minutus* migrate westwards during autumn from Europe

Fig. 8. (*a*) Schematic pattern of wind and pressure zones at earth's surface. (*b*)
Great shearwaters *Puffinus gravis* migrate from their breeding colonies in the
South Atlantic by a clockwise movement in the North Atlantic, thereby
benefitting from following winds in the zones of trade winds as well as in the
northern west-wind zone. Schematic pattern of migration based on Voous &
Wattel (1963) and Cramp & Simmons (1977).

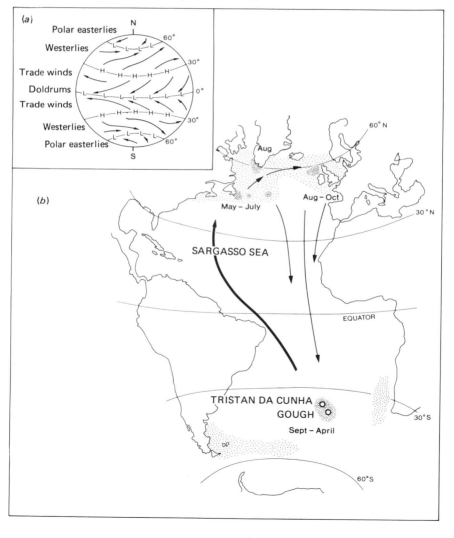

in the west-wind zone across the North Atlantic to exploit the waters off New-foundland, which are so rich in food (Coulson, 1966; Tuck, 1971).

Geographical wind patterns influence the evolution of migratory pathways also in land birds, although the effect is less striking than in seabirds. British summer residents leave in autumn on SSE to SE courses, later to shift over continental Europe towards SW and Iberia. This dog-leg route from Britain to Iberia is prob-ably more advantageous than the straight route because the birds benefit from the winds. They take advantage of the southwest to northwest winds frequently pre-vailing in Britain. The migrants could travel even faster if they departed from Britain further east than southeast, but this advantage is counteracted by the negative effect of an excessively long detour to the destination areas in Iberia and Africa (Evans, 1966).

Similarly, the route of birds migrating over the North Atlantic Ocean from North America to the West Indies and South America seems to be an optimal compromise between maximal benefit from winds and minimal detour. The birds depart from northeastern North America towards southeast, taking advantage of the strong northwesterly winds after passages of cold fronts in this area, and their flight direction gradually shifts towards south or even south southwest when they fly into the zone of easterly trade winds (Richardson, 1976; Williams *et al.*, 1977).

An exceptionally profound influence of wind on the migratory pattern is found in the redwing thrush *Turdus iliacus*. In this species wind seems to affect not only the route of migration but also the birds' destination (Fig. 9). During late autumn the redwing is one of the most common species migrating across the North Sea, from Norway to Britain and continental Europe. Large numbers of redwings also depart southwestwards from the countries on both sides of the Baltic Sea, to winter quarters in west and southwest Europe. According to radar studies in both Britain and Scandinavia all these southwesterly movements usu-ally take place under following northerly to easterly winds and fair weather.

However, radar observations in South Scandinavia (Alerstam, 1975b) also show that soon after the passage of a cold front, mass movements towards south-east occur in brisk northwesterly winds. A large proportion of the thrushes departs from Norway, and, provided the wind remains favourable, the birds migrate not only during the night, but continue during the morning in full day-light, when hundreds of thousands of migrating redwings have been counted by field observers in southeasternmost Sweden. Due to the strong tailwinds, the ground speed of the redwings flying southeast is on the average 22 m s^{-1}, and on some occasions up to 28 m s^{-1}, more than twice the birds' air speed.

The courses of these redwings point towards the region around the Caspian, Black and easternmost Mediterranean Seas, and there are a few winter recoveries from this general area of redwings ringed in Scandinavia, although the vast majority of winter ringing recoveries come from West Europe. However, ringing data cannot be used for comparing the quantitative importances of different winter areas, and certainly the region to the southeast is relatively a more important redwing winter region than indicated by numbers of ringing recoveries. Ringing data from the east coast of the Baltic Sea confirm that redwing migrants in this region travel either to SW Europe or to the Black and Caspian Sea regions.

The most illuminating ringing results come from Britain, where quite a number of redwings ringed in winter have been recovered during a subsequent winter in the Near East and southern Russia (Fig. 9). This indicates that winds prevailing when the redwings are prepared to depart from Scandinavia determine which winter quarter will be visited – the birds take advantage of following northeast

Figure 9. Autumn migration of redwing *Turdus iliacus* in Europe. Under northerly and easterly winds birds depart from Scandinavia and the Baltic region towards southwest to winter quarters in Britain and west/southwest Europe. With strong winds from the west and northwest behind cold fronts, they depart southeast towards the region around Caspian, Black and eastern Mediterranean Seas. Winter recoveries in this latter region of birds ringed in a previous winter in Britain (Mead, 1976) are indicated on the map (circles). Vertical lines indicate the breeding range, the stippled area, the winter range; lines and stippling, breeding and winter range.

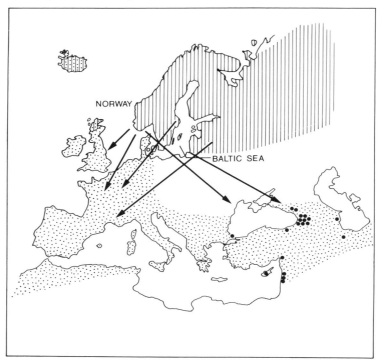

or northwest winds to travel swiftly either to Britain/west Europe or the Near East/southeast Europe, respectively. So, an individual redwing may winter in Britain one year and in Soviet Georgia the next.

In general, migrating birds are faithful to their winter as well as to their breeding sites. The exceptional flexibility in the redwing brings the advantage that energy costs for the migratory journey are minimized.

Weather and timing of migratory flights

Migratory intensity varies dramatically between different days and nights, even during the peak period of migration. This variation is due mainly to weather. During recent decades radar has been used extensively to monitor migratory intensity in relation to weather in the north temperate zone, while next-to-nothing is known about this relationship in other climatic regions.

The occurrence of intensive spring and autumn migration in relation to synoptic weather features is shown in Fig. 10. Dense spring migration usually takes place with warm southerly winds, to the west of a high pressure and to the east of a low pressure centre. Furthermore, it occurs in the moist, warm sector of a depression, and spring migrants heading northeast also depart during westerly or southwesterly winds behind cold fronts of relatively mature cyclones. In contrast, intensive autumn migration takes place with cold northerly winds to the east of a high pressure centre and west of a cyclone. Migrants on their way towards the southeast depart under brisk northwesterly winds immediately after a cold-front passage while southwesterly migrants often wait for one or two further days, and depart when the anticyclone has moved closer and winds blow from the north or northeast. Spring as well as autumn migration occurs under anticyclonic conditions with calm and fair weather (Richardson, 1978).

In many studies, migratory intensity has been correlated with a great number of different weather variables, and multivariate statistical analyses have been used to investigate which combination is best for predicting migration. Most studies agree that wind and precipitation are key factors: dense migration takes place with following winds and no precipitation. Once these factors are taken into account predictions of migratory intensity do not become appreciably better by considering additional weather variables. This is equivalent to saying that under constant wind and precipitation conditions, only weak correlations exist between migratory intensity and other weather factors.

It is easy to see the selection pressures behind the close association of intensive migration with following winds. Flying time and energy costs for the migratory journey become greatly reduced for birds using tailwinds to achieve a high ground speed. It is advantageous for birds to avoid the stresses of increased heat loss caused by wetting of plumage, and possibly of disorientation, during flight under conditions of precipitation.

Migrating birds' responses to weather may have evolved not only in response to selection pressure for favourable flight conditions but also for arrival at their destinations under favourable living conditions and for departure when conditions at the point of origin worsen. The dominant influence of following winds on migratory intensity, as found in most studies, indicates that flight conditions

Figure 10. Schematic illustration of synoptic weather situations associated with spring (upper) and autumn (lower) migration in the Northern Hemisphere. Stripled areas show situations in relation to pressure cells, warm and cold fronts, in which dense bird migration often takes place. Thin lines indicate isobars. Winds blow approximately parallel to the isobars, clockwise around a high-pressure centre and anticlockwise around a low-pressure system, and wind speed is inversely proportional to spacing between isobars. Based on Richardson (1978).

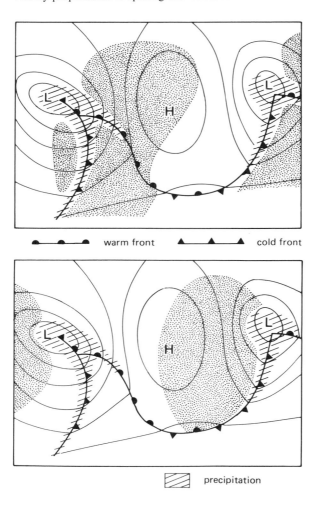

generally constitute the primary selecting force for the short-term effect of weather on bird migratory activity, while living conditions at departure or destination areas play at most a subsidiary role. However, there are a few interesting exceptions.

In May, intensive northeastward songbird migration takes place in New England. Departures from Massachusetts of birds reaching Maine after a night's migration were observed on radar by Nisbet & Drury (1968). A combination of onshore crosswinds, low but rising humidity, rising temperature, low pressure and absence of rain is most reliable for predicting heavy migration. The synoptic interpretation is that 'migration is concentrated into the zone to the west of the cool dry centre of a high-pressure cell and east of a warm trough of low pressure. This zone usually marks a strong northward flow of warm air; the flow is usually less strong than in the trough to the west, but is more likely to persist and less liable to develop into cyclonic disturbances' (Nisbet & Drury, 1968). As a result, the migrants usually encounter high temperatures and avoid rain in their destination area the next morning. Indeed, the above-mentioned combination of weather variables in the departure area is remarkably good for predicting warm and dry weather at the destination next day. Any reinforced response by the departing birds to tailwinds *per se* would increase the risk of encountering cold and rainy weather after arrival in the northerly areas.

Hence in this case natural selection has made the migrants masters in the task of forecasting weather the next day in areas hundreds of kilometres to the northeast. Selection for favourable living conditions at the destination, warm, dry weather and therefore plenty of available food for the insectivorous birds, overrides the narrow selection for maximal tailwinds en route.

For late-autumn migrants, like pigeons, crows, starlings and finches, a falling temperature is positively correlated with migratory intensity even when the effect of wind is taken into account. Hence, in these migrants living conditions at the take-off site exert a significant selection pressure on the migrants' responses to weather.

During winter and early spring, cold, snowfall and frost may release hard-weather movements or retreat migration.

The use of radar for monitoring bird migration brings two major drawbacks: different species usually cannot be distinguished and migration at extremely low altitudes is often missed because it is below radar coverage. Direct field observations of visible bird migration do not have these drawbacks, and furnish an important complement to radar data. The most serious sources of bias in field studies of visible migration are that high-altitude movements of small birds escape notice, and that migration along leading-lines is overemphasized.

Field observations reveal considerable differences between responses to weather by different species. Weather characteristic of maximal visible autumn

Fig. 11. Weather associated with intensive visible autumn migration of wood pigeons and chaffinches/bramblings (these two species travel in mixed flocks and cannot be counted separately) at Falsterbo, South Sweden. Average numbers of migrants are calculated for the seasonal peak period of migration. Data from Roos (1977, 1978*b*) and Alerstam (1978).

WOOD PIGEON	BRAMBLING/CHAFFINCH
Light and moderate NE winds	Moderate SW winds
Clear sky with scattered cumulus clouds	
No rain, low humidity and good visibility	No rain
High and rising barometric pressure	
Low and sinking temperature	Mild but sinking minimum temperature
Average no. of migrating wood pigeons	Average no. of migrating chaffinches/bramblings
on days with NE winds: 8900	on days with NE winds: 1500
on days with SW winds: 1600	on days with SW winds: 120500

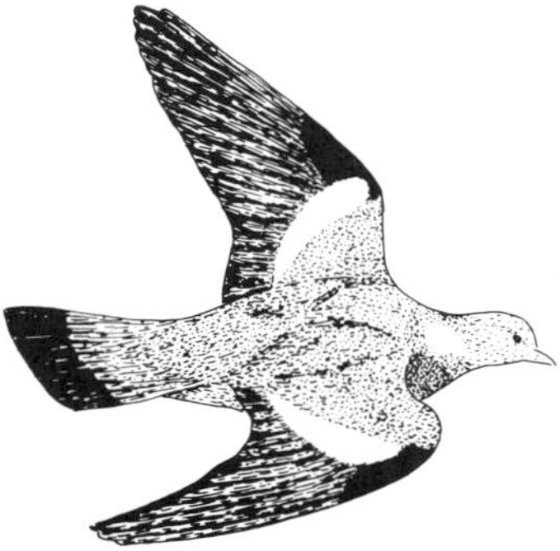

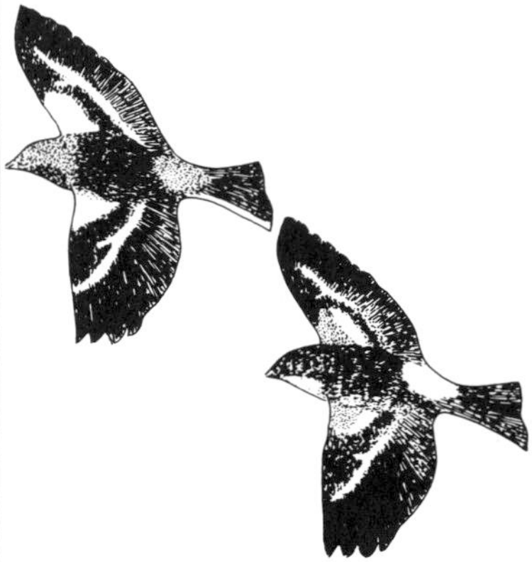

migration of wood pigeons and finches (chaffinches *Fringilla coelebs* and bramblings *F. montifringilla*) at Falsterbo in South Sweden are compared in Fig. 11. Wood pigeons are relatively choosy with regard to following winds, and prefer to migrate after passage of a cold front, under cold northerly winds and clear skies. In contrast, almost a hundred times as many finches are counted by the field observer under headwinds as compared to tailwinds. Rain suppresses finch migration, but there are no significant correlations with cloudiness, visibility, humidity, or barometric pressure.

Radar, in combination with observations from aircraft and with vertical telescopes, reveals that great numbers of finches migrate also with following winds, under weather conditions similar to those associated with intensive migration of wood pigeons. Under such conditions the finches fly on a broad front at relatively high altitude, from several hundred up to one or two thousand metres, above the detection range of a field observer. Hence, the large number of finches registered by the field observer under headwinds does not mean that the finches *prefer* headwinds for their migration, but rather that circumstances force them to take flight under these energetically unfavourable conditions also.

The adaptive values of the different responses to weather can be explained as follows: weather in northwest Europe is characterized by the eastward flow of successive cyclones. Sometimes an intervening anticyclone builds up and there are no cyclones passing for some days. In the east and central part of the high-pressure cell, conditions are optimal for autumn migration, with following winds and fair weather. Migrants of all sorts, diurnal and nocturnal, finches and pigeons, seize this opportunity and depart in vast numbers. Migration takes place on a broad front at relatively high altitudes, where the following winds are strongest (Fig. 12(*a*)). However, especially in late autumn, anticyclonic interludes may seldom occur, so that cyclones pass in rapid unbroken succession. Looking through the daily weather maps for some years, I found that in the southern North Sea area frontal passages occur on the average every third day between mid September and mid November. Regularly there are periods with families of frontal depressions, with fronts passing on five to seven out of ten days. During such periods there are no better occasions for migration than the brief wake of a cyclone just having passed eastwards, before the arrival of the next depression (Fig. 12(*b*)). Finches depart on these occasions, since they would otherwise be delayed for long periods waiting for anticyclones and following winds. In order to minimize the adverse effect of headwinds they fly low over the ground and follow coastlines.

Chaffinches carry out their migratory journey from Scandinavia to Britain or France via the Low Countries by a great number of short hops, each consisting of a few hours' flight in the morning. This seems to be a reasonable strategy in the extremely windy and variable late autumn weather of this region. Scandina-

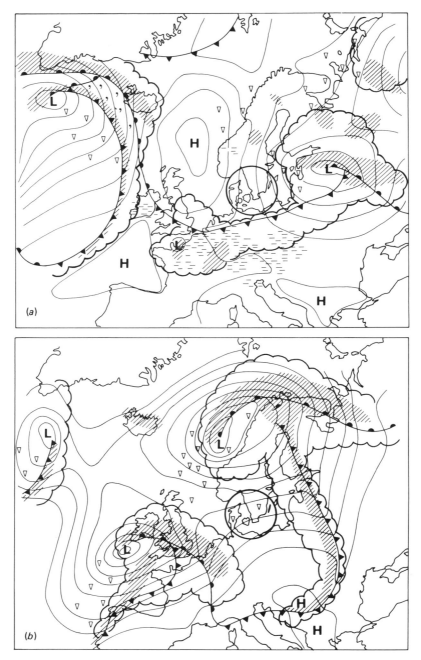

Figure 12. Weather is optimal for autumn migration during an anticyclonic interlude between passages of depressions, under northerly winds and clear

vian wood pigeons destined for France and the Iberian peninsula fly at high speed (Table 1) and travel a relatively long distance during a morning's flight. They can therefore better afford to wait for anticyclones bringing tailwinds. However, this strategy leads to delays in pigeon migration in some autumns, and if tailwinds fail for a long time the pigeons finally give up waiting and depart at low altitude under crosswinds, a compact cloud cover and poor visibility (Roos, 1978*b*). It is a rule among migrating birds that the threshold in favourable weather for the release of migratory activity decreases as the delay since the previous instance of migration grows longer.

When migrants encounter adverse weather like headwinds, rain or fog in flight they land as soon as possible. The occurrence of sizeable falls of night migrants at the English east coast in the autumn is usually associated with favourable departure weather in Scandinavia and adverse weather over the North Sea off the English coast (Lack & Parslow, 1962).

Migrants crossing the Gulf of Mexico in spring normally do not land immediately upon reaching the North American coast but continue 50–150 km across marsh and prairie to the large inland forests before descending. However, when the trans-Gulf migrants meet rain or opposing winds over the Gulf or the North American coast they become grounded, sometimes in tremendous concentrations, on the first available land, barrier islands, marshland and isolated coastal woodlands (Lowery, 1945; Gauthreaux, 1971).

In late summer, arctic waders often fly non-stop across Scandinavia to the shallow North Sea shores. They probably take off from the White Sea region. In years with a lot of northerly and easterly winds the numbers of such waders resting in South Scandinavia are particularly low. It is only in summers with frequent cyclonic passages, brisk southwesterly winds, cool and rainy weather that they occur in large numbers, and many can even be seen on low-altitude coastbound migration into the headwinds. Obviously the waders interrupt their long non-stop flight in years with adverse weather (Roos, 1975).

Caption to Fig. 12 (cont.).
skies after a cold-front passage. On occasion (*a*), radars in southern Scandinavia (encircled area) demonstrated high-altitude mass migration of passerine as well as non-passerine birds during both light and dark hours of the day, and field observers detected the intensive migration of larger birds like buzzards and wood pigeons. During periods with passages of cyclones in rapid succession, finch migrations slip between the cyclones in moderate southwest winds, as exemplified in (*b*). On this occasion several hundred thousands of finches migrating at low altitude were counted by the Falsterbo field observer, while according to radar, there was insignificant migration higher up.

Flight altitude

The first radar studies of bird migration revealed that great numbers of birds regularly fly at much greater altitude than had been generally assumed, far above detection range of a field observer. Technical limits of radars as well as local environmental effects, leading to poor radar coverage at low or high altitudes, make it difficult to attain a reasonably complete quantitative height distribution of bird migration. However, this has been accomplished in a few studies, in Britain (Eastwood & Rider, 1965), Switzerland (Bruderer, 1971), Massachusetts (Nisbet, 1963b) and Louisiana (Able, 1970). In addition, the height distribution of night migrants over Illinois and Nebraska has been mapped by flying altitudinal transects with a light aircraft equipped with extra lights (Bellrose, 1971). According to these studies, primarily concerned with passerine night migration, 50% of all migrants fly below 400–700 m above ground level and 90% below 1000–2000 m. In the most extensive and detailed of these studies, night migrants were recorded with a tracking radar over the Swiss lowlands in spring, and 50% of all birds travelled below 700 m and 90% below 2000 m (Bruderer, 1971). Maximum height in various radar studies lies between 3000 and 6300 m.

Passerines depart on night migration 30–40 min after sunset, and the mean height rapidly increases to a maximum in the early night, thereafter slowly decreasing as migratory intensity fades. Migration during daytime generally occurs at slightly lower altitudes than during the night.

Under special circumstances, the mean height distribution of bird migration may be quite different from that described above. Spring migrants arrive at high altitude over Louisiana after crossing the Gulf of Mexico; about three quarters of them fly between 1000 and 3000 m (Gauthreaux, 1972). Shorebirds departing across the West Atlantic from Nova Scotia in autumn have an overall median altitude at 1700 m, and 10% fly above 3900 m with the highest flock having been detected at 6650 m (Richardson, 1979). Migration across the West Atlantic to a large extent proceeds at altitudes between 1000 and 5000 m as demonstrated by radar studies at Bermuda, Puerto Rico and the Lesser Antilles (Richardson, 1976; Williams et al., 1977). Especially over Puerto Rico (mainly passerine migration) and over Antigua and Barbados (mainly shorebird migration) heights are very great, and the medians may reach 4000–5000 m. Some migrants have been detected as high as 6800 m over Puerto Rico. In contrast, over Tobago close to the South American mainland, altitudes are mostly below 500 m and the birds are obviously preparing to land.

Which factors govern the birds' choice of flight altitude? There are two general incentives for the birds to fly high. One is that optimum flying speed (V_{mr}) is inversely related to air density so that V_{mr} in the standard atmosphere increases by about 5% per thousand metres. A high flying speed *per se* may be advantageous to the migrants. Secondly, the power required to fly at this speed, P_{mr}, also increases with increasing altitude, but at a rate slightly lower than that of flight speed, leading to a reduction in the cost of transport with increasing altitude. The relatively low rate of increase in P_{mr} with height is due mainly to the assumption that basal metabolism and maintenance power are constant irrespective of flying height. Hence, heat from the flight muscle work is assumed to compensate for low ambient temperatures at high altitudes. However, the reduction in cost of transport with increasing altitude is very modest, about 1% per thousand metres in small birds which have a relatively large fraction of maintenance power in their total flight power budget.

According to these arguments the optimum height for a migrant to fly is the highest at which it can get just sufficient oxygen to maintain P_{mr}. As fuel is used up in the course of a long flight, and the bird's weight decreases, P_{mr} gradually decreases at any particular height. The bird can then maintain P_{mr} at a progressively higher altitude as its flying distance increases. Hence, the best strategy for a migrant is to climb gradually during its flight, always flying at the greatest height at which oxygen pressure is sufficient to maintain P_{mr}. This strategy is called cruise climb (Pennycuick, 1975, 1978), and leads to the prediction that a small bird should increase its height by about 1300 m over an air distance of 1000 km. The cruise-climb strategy has been suggested as an explanation of the high altitudes of migrants having travelled long distances non-stop over the Gulf of Mexico and the West Atlantic.

However, I think it is doubtful whether the birds ever use the cruise-climb strategy, as defined above, because there is an important agent that profoundly affects birds' flight speeds and costs of transport at different altitudes: wind. Wind speed is often of the same order of magnitude as the birds' air speed, and winds vary considerably at different heights. One should therefore expect that birds prefer to fly at altitudes with the most favourable winds. Indeed, there is much evidence from radar studies that wind is a key factor in explaining the height distribution of bird migration.

Due to friction, wind speed usually is less at low than at high altitude. As a consequence, migratory birds fly relatively low under headwinds and high with following winds. The difference in height distribution of bird migration under opposing as compared to following winds is exemplified in Fig. 13 (a) and (b). Migrants are concentrated in strata with minimal headwinds and maximal tailwinds. Even the high flights over Puerto Rico may be explained with reference to wind (Fig. 13 (c)). In this region, easterly trade winds blow at right angles to the birds' goal direction. However, at high altitudes the trade winds are less consistent and strong than they are closer to surface, and consequently the birds will increase their ground speed and reduce their energy costs by flying at these great heights.

Migrants with widely different goal directions frequently travel simultaneously, and if winds shift markedly between altitudes, they will select different flying heights (Steidinger, 1972; Bruderer, 1975). One example in Fig. 14 shows such an instance, where migrants with goal directions towards ENE exploit the strong westerly winds at relatively high altitude, while northbound migrants

Fig. 13. Height distribution of bird migration. (*a*), (*b*) Two nights in spring over the Swiss lowlands with opposed and following winds, respectively. Both nights were clear and the 0°C-temperature level was at 2000 m altitude (from Bruderer, 1971). (*c*) Passerine migration approaching Puerto Rico from the north on one October morning. The birds have travelled non-stop 2000–3000 km over the West Atlantic before reaching Puerto Rico, where easterly trade winds blow at right angles to the migrants' track directions towards the south. The 0°C-level is about 5000 m (from Richardson, 1976).

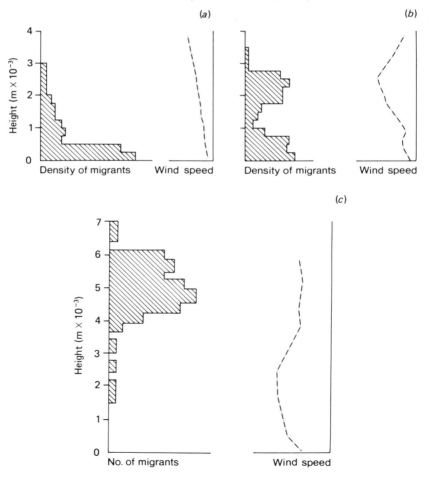

travel at lower levels, where they find their best winds. Indeed, birds have an impressive capacity for finding the heights with the most favourable winds, but how they manage to do this is not known.

As a rule, migrating birds avoid flying in clouds, and during full overcast with a low cloud ceiling, they become concentrated below the clouds. Birds generally migrate above banks of fog, and concentrations of migrants sometimes occur immediately above and below low clouds. There are radar records of birds migrating in cloud layers demonstrating that they can maintain a well oriented flight under these conditions, at least for a short time (Griffin, 1973). Disoriented movements, where birds fly in directions which vary at random and change irregularly from minute to minute, have been revealed by radar under special weather conditions with full overcast, fog and rain.

There is a particularly nice demonstration of the effect of clouds on the flying height of swifts. Altimeters were pasted to the backs of the birds before they were released at various distances from their breeding sites, where birds as well as altimeters were recovered after the homing flights. The altimeter is based on the principle that the range in air of α-particles from a radioactive substance is inversely proportional to air density, so that after development of a photographic emulsion with the α-particles recorded, one can find the bird's maximum altitude and also for how long the bird had been flying in different height intervals. Under

Fig. 14. Directions and speeds of migrating birds (left) and winds (right) on a spring night over the Swiss lowlands. Birds with different goal directions select different flight altitudes, with the most favourable winds (from Steidinger, 1972).

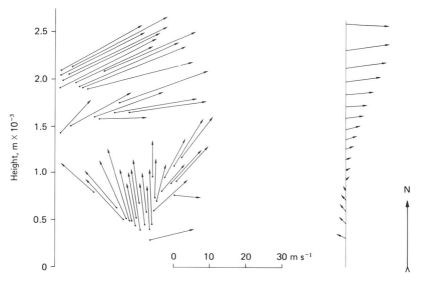

clear skies the average maximum flying height of the swifts was 2300 m, and ten out of fifteen birds flew above 3000 m, the highest at 3600 m. In contrast, with overcast conditions the average maximum altitude was only 700 m, ranging from less than 50 m up to 1700 m in nineteen different birds (Gustafson, Lindkvist & Kristiansson, 1973; Gustafson, Lindkvist, Gotborn & Gyllin, 1977).

Low temperatures seem not to prevent birds from migrating at great heights. The highest migrants over Puerto Rico experienced temperatures of about $-12°C$, and over Switzerland migration is perfectly regular at heights with temperatures around -10 to $-15°C$. In fact, the birds' capacity to fly under conditions of low temperature and low oxygen pressure probably is so great that it is hard to believe: on 9 December 1967 a radar controller in Northern Ireland reported an echo at high altitude moving south over the Hebrides. The radar height finder indicated an altitude between 8000 and 8500 m. The pilot of an aircraft in the vicinity was asked by radio to make a course deviation to pass near the position of the echo. In doing so, the pilot reported a flock of about 30 swans, at just over 8200 m. The radar controller had the echo in sight until it disappeared at the North Irish coast, when the flock had probably descended below radar coverage. The observation probably refers to whooper swans, which are known to migrate, sometimes in the middle of winter, from Iceland to Britain. At 8200 m the swans were at the tropopause, clear of large snow showers over the sea. Temperature was $-48°C$ and very strong northerly tailwinds about 50 m s^{-1} blew at this altitude (Stewart, 1978; Elkins, 1979).

A colleague of mine travelling between Geneva and Copenhagen saw two small flocks close behind each other (about five and three individuals) of curlew-like birds on 19 June 1974. The aircraft captain confirmed the observation and informed her that the position was immediately north of Frankfurt am Main and the height 10 000 m above sea level. (B. Lindkvist, personal communication).

These observations indeed are extraordinary. Are such enormous flying heights merely exceptional, or do some birds regularly migrate that high?

Flight direction and wind

Do migrating birds drift by the winds or do they compensate for wind drift (Fig. 15)?

At first thought, one would expect radar to be an ideal instrument for easily solving this problem. If the mean track direction of bird migration is shifted to the right on occasions with winds from the left, and to the left with winds from the right, this seems good evidence that birds are subjected to wind drift. In fact, almost all radar studies reveal that the overall mean track direction of bird migration is shifted in relation to wind in this manner. However, this may also be explained by a phenomenon called pseudodrift: different migrants are aloft under different wind conditions (Fig. 16).

Pseudodrift may occur because different species prefer different winds for their migratory flights, but also because populations of the same species differ in their choice of winds during migration. Such intraspecific pseudodrift has been confirmed in the eider duck and wood pigeon in South Scandinavia. By way of example, eiders departing eastsoutheast in spring prefer westerly and northwesterly winds, while eiders travelling eastnortheast fly in greatest numbers under southwesterly winds. Hence, the mean track direction of eider migration varies according to which population migrates in largest numbers.

By carefully distinguishing radar echoes from different cohorts of migrants, one may eliminate a large part of the effect of pseudodrift. A few radar studies over land and coastal regions indicate that, after pseudodrift is eliminated in this way, track directions of migrants are constant irrespective of wind – in other words, the birds compensate completely for wind drift. However, other studies

Fig. 15. A bird's track velocity T (flight speed and direction in relation to ground) is the sum of its heading velocity H (speed and direction in relation to air) and wind velocity W. (a) Birds flying on a constant heading towards their goal will be drifted by the winds. (b) In order to maintain a constant track direction irrespective of wind, birds must compensate for wind drift by heading into the wind.

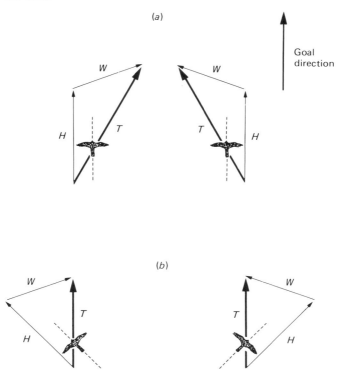

under similar conditions indicate that a drift effect remains even after elimination of pseudodrift.

All radar studies agree that true wind drift occurs over the open sea. In South Scandinavia, wood pigeons and cranes compensate completely for wind drift over land, but drift when they migrate across the Baltic Sea. The variation in track directions of birds flying over the sea is, however, less than expected if they drift from a fixed heading. Perhaps migrants over the sea, devoid of landmarks for orientation to compensate wind drift, use the moving sea waves for this purpose, leading to partial drift (Alerstam & Pettersson, 1976)?

Hence, in spite of two decades of radar studies any final solution to the wind drift/compensation problem is not yet in prospect. The issue obviously is much more complex than first suspected: in some instances migrating birds do compensate completely for wind drift over land and coastal areas, but in other instances they are probably drifted to some extent by wind, and over the open sea they fail to compensate completely.

Could it be advantageous for migrants to allow themselves to be wind drifted under certain conditions?

(1) Drift will always be disadvantageous if winds blow from a constant direction during the migratory journey. Under these conditions drift leads to a long flight path and flying time to the goal, even if the migrants continually readjust their heading direction towards their goal.

Fig. 16. Vectors show the ground speed of birds travelling along different track directions in a migratory movement with a total spread of tracks at 100°. Under calm conditions, ground speed of migrants will be equal irrespective of their track, whereas, under cross-winds, ground speed will be low in birds travelling into the wind and high in birds flying on tracks with following winds. As birds generally prefer favourable winds, giving a high ground speed, for their migratory flights, migrants with preferred track directions pointing into the wind will depart in smaller numbers than those flying on tracks with following winds. As a result, the mean track direction of the migratory movement will shift in relation to wind. Since this effect is independent of the individual birds' compensation for or drift by winds, it is called pseudodrift.

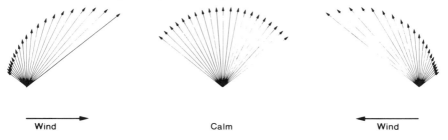

Wind   Calm   Wind

The fastest way to reach the goal is to compensate for wind drift and fly along the straight-line track to the destination.

(2) However, if winds vary during the migratory journey, the birds can reduce flying time, and costs of transport, if they allow themselves to be drifted by winds from the left during one flight, and wait for favourable winds from the right during subsequent flights. Hence, if winds shift significantly between the separate flights of the migratory journey, the birds should minimize the distance to their destination after each flight, and their optimum strategy is to allow extensive drift far away from their destination and gradually compensate to a higher and higher degree on approaching it, finally compensating completely during the last flight.

(3) If the wind pattern shifts in a predictable way along the migratory route, so that wind from the right is succeeded by wind of a similar net force from the left, or vice versa, the birds will minimize flying time by travelling on a rather constant heading, allowing themselves to be drifted by wind. Perhaps this is what migrants flying from North to South America over the West Atlantic do. Their heading direction is towards southeast to southsoutheast, both off New England, where they drift eastwards when departing under strong westerly to northwesterly winds after passages of cold fronts, and over the Caribbean Sea, where they drift westwards in the zone of easterly trade winds (Williams & Williams, 1978).

(4) Migrating birds may also save flying time by allowing themselves to be partially drifted by strong winds at high altitude and correcting for the displacement by overcompensating at low altitude under relatively weaker winds. In fact, overcompensation in the sense that the migrants' track directions, when comparing occasions with different winds, are shifted into the winds, has been found in low-altitude visible migration of, e.g., chaffinches and starlings.

In the light of these various possibilities it seems not so odd that a simple answer to the question of whether or not migrants compensate for wind drift is lacking. I think there remains a lot to learn about the migrants' exploitation of winds, involving different strategies in different wind conditions, to achieve a swift and safe migratory journey at small energy costs.

Why do migrating birds fly along coastlines? Bird migration along coastlines sometimes leads to spectacular concentrations of migrants at certain peninsulas, many of which are resorts of bird-watchers' pilgrimages. Migrants' coasting behaviour probably is part of their strategy to get the better of winds. Coastal migration occurs mainly under opposed and cross-winds, while migrants usually fly across the coast and depart over the open sea with following winds. Due to differences in friction, winds generally are stronger over the sea than over land. Migrants minimize the headwind force by following the coast, where they also can use local topography and vegetation to gain additional protection from the wind. Furthermore, over the sea they will be exposed to wind drift, and under certain cross-winds it is beneficial to follow a coastline some distance goalwards rather than to depart directly over the sea.

Final remarks

The key word when considering the course and timing of bird migration is adaptation. The adaptation of migrating birds' behaviour to hazards and hardships in the environment reaches a level of precision that is highly impressive to the most skilful and experienced human pilot. Much remains to be unravelled about the amazing performances by billions of birds, travelling across continents, deserts, oceans and glaciers in their striving for food, suitable breeding grounds and safety from predators.

This chapter was prepared with support from the Swedish Natural Science Research Council (grant B 0141–101). I thank Inga Rudebeck for typing the manuscript and Astrid Ulfstrand for drawing the figures.

References

Able, K. P. (1970). A radar study of the altitude of nocturnal passerine migration. *Bird-Banding*, **41**, 282–90.

Alerstam, T. (1975a). Crane *Grus grus* migration over sea and land. *Ibis*, **117**, 489–95.

Alerstam, T. (1975b). Redwing *Turdus iliacus* migration towards southeast over southern Sweden. *Vogelwarte*, **28**, 2–17.

Alerstam, T. (1976). *Bird Migration in Relation to Wind and Topography*. Ph.D. thesis, University of Lund.

Alerstam, T. (1978). Analysis and a theory of visible bird migration. *Oikos*, **30**, 273–349.

Alerstam, T. & Pettersson, S-G. (1976). Do birds use waves for orientation when migrating across the sea? *Nature*, **259**, 205–7.

Bellrose, F. (1967). Radar in orientation research. *Proceedings from 14th International Ornithological Congress*, pp. 281–309. Oxford: Blackwell.

Bellrose, F. (1971). The distribution of nocturnal migrants in the air space. *Auk*, **88**, 397–424.

Bernis, F. (1975). (Migration of Falconiformes and *Ciconia* spp. through the straits of Gibraltar.) *Ardeola*, **21**Especial, 489–594 (Spanish, English summary).

Broekhuysen, G. J. & Siegfried, W. R. (1971). Dimensions and weight of the steppe buzzard in southern Africa. *Ostrich*, Suppl. **9**, 31–9.

Bruderer, B. (1971). Radarbeobachtungen über den Frühlingszug im Schweizerischen Mittelland. *Ornithologische Beobachter*, **68**, 89–158.

Bruderer, B. (1975). Zeitliche und räumliche Unterschiede in der Richtung und Richtungsstreuung des Vogelzuges im Schweizerischen Mittelland. *Ornithologische Beobachter*, **72**, 169–79.

Bruderer, B. & Weitnauer, E. (1972). Radarbeobachtungen über Zug und Nachtflüge des Mauerseglers *Apus apus. Revue Suisse de Zoologie*, **79**, 1190–200.

Coulson, J. C. (1966). The movements of the kittiwake. *Bird Study*, **13**, 107–15.

Cramp, S. & Simmons, K. E. L. (eds) (1977). *The Birds of the Western Palearctic*, vol. 1. Oxford: Oxford University Press.

Devillers, P. (1977). The skuas of the North American Pacific coast. *Auk*, **94**, 417–29.

Dolnik, V. R. & Gavrilov, V. M. (1973). Energy metabolism during flight of some passerines. In *Bird migrations. Ecological and Physiological Factors*, ed. B. E. Bykhovskii, pp. 288–96. Jerusalem: John Wiley.

Dorst, J. (1962). *The Migrations of Birds*. London: Heinemann.

Eastwood, E. & Rider, G. C. (1965). Some radar measurements of the altitude of bird flight. *British Birds*, **58**, 393–426.

Elkins, N. (1979). High altitude flight by swans. *British Birds*, **72**, 238–9.

Elliott, C. C. H., Waltner, M., Underhill, L. G., Pringle, J. S. & Dick, W. J. A. (1976). The migration system of the curlew sandpiper *Calidris ferruginea* in Africa. *Ostrich*, **47**, 191–213.

Emlen, S. T. (1974). Problems in identifying bird species by radar signature analyses: intra-specific variability. In *A Conference on the Biological Aspects of the Bird/Aircraft Collision Problem*, ed. S. A. Gauthreaux, pp. 509–24. Clemson University.

Emlen, S. T. & Demong, N. J. (1978). Orientation strategies used by free-flying bird migrants: a radar tracking study. In *Animal Migration, Navigation, and Homing*, ed. K. Schmidt–Koenig & W. T. Keeton, pp. 283–93. Berlin: Springer–Verlag.

Evans, P. R. (1966). Migration and orientation of passerine night migrants in northeast England. *Journal of Zoology, London*, **150**, 319–69.

Fry, C. H., Ash, J. S. & Ferguson-Lees, I. J. (1970). Spring weights of some Palaearctic migrants at Lake Chad. *Ibis*, **112**, 58–82.

Gauthreaux, S. A. (1971). A radar and direct visual study of passerine spring migration in southern Louisiana. *Auk*, **88**, 343–65.

Gauthreaux, S. A. (1972). Behavioral responses of migrating birds to daylight and darkness: a radar and direct visual study. *Wilson Bulletin*, **84**, 136–48.

Gladwin, T. W. (1963). Increases in the weights of Acrocephali. *Bird Migration*, **2**, 319–24.

Glutz von Blotzheim, U. N., Bauer, K. M. & Bezzel, E. (1971). *Handbuch der Vögel Mitteleuropas*, vol. **4.** Frankfurt am Main: Akademische Verlagsgesellschaft.

Glutz von Blotzheim, U. N., Bauer, K. M. & Bezzel, E. (1975). *Handbuch der Vögel Mitteleuropas*, vol. **6.** Wiesbaden: Akademische Verlagsgesellschaft.

Greenewalt, C. H. (1975). The flight of birds. *Transactions of the American Philosophical Society*, New Series, vol. **65,** part 4.

Griffin, D. R. (1973). Oriented bird migration in or between opaque cloud layers. *Proceedings of the American Philosophical Society*, **117**, 117–41.

Gustafson, T., Lindkvist, B. & Kristiansson, K. (1973). New method for measuring the flight altitude of birds. *Nature*, **244**, 112–13.

Gustafson, T., Lindkvist, B., Gotborn, L. & Gyllin, R. (1977). Altitudes and flight times for swifts *Apus apus. Ornis Scandinavica*, **8**, 87–95.

Hart, J. S. & Berger, M. (1972). Energetics, water economy and temperature regulation during flight. *Proceedings from the 15th International Ornithological Congress*, pp. 189–99. Leiden: Brill.

Hussel, D. J. (1969). Weight loss of birds during nocturnal migration. *Auk*, **86**, 75–83.

Lack, D. (1959). Migration across the sea. *Ibis*, **101**, 374–99.

Lack, D. (1959–1963). Migration across the North Sea studied by radar. Parts 1–4. *Ibis*, **101**, 209–34; **102**, 26–57; **104**, 74–85; **105**, 1–54 and 461–92.

Lack, D. & Parslow, J. L. F. (1962). Falls of night migrants on the English east coast in autumn 1960 and 1961. *Bird Migration*, **2**, 187–201.

Larkin, R. P., Griffin, D. R., Torre-Bueno, J. R. & Teal, J. (1979). Radar observations of bird migration over the western North Atlantic Ocean. *Behavioral Ecology and Sociobiology*, **4**, 225–64.

Lowery, G. H. (1945). Trans-Gulf spring migration of birds and the coastal hiatus. *Wilson Bulletin*, **57**, 92–121.

McNeil, R. & Burton, J. (1977). Southbound migration of shorebirds from the Gulf of St. Lawrence. *Wilson Bulletin*, **89**, 167–71.

Mascher, J. W. (1966). Weight variation in resting dunlins *Calidris aplina* on autumn migration in Sweden. *Bird-Banding*, **37**, 1–34.

Mead, C. (1976). In *BTO-news*, ed. D. Glue, No. 80, p. 3. Tring: The British Trust for Ornithology.

Moreau, R. E. (1972). *The Palaearctic–African Bird Migration Systems*. London: Academic Press.

Morrison, R. I. G. (1977). Migration of arctic waders wintering in Europe. *Polar Record*, **18**, 475–86.

Nisbet, I. C. T. (1963*a*). Weight-loss during migration, Part 2. *Bird-Banding*, **34**, 139–59.

Nisbet, I. C. T. (1963*b*). Measurements with radar of the height of nocturnal migration over Cape Cod, Massachusetts. *Bird-Banding*, **34**, 57–67.

Nisbet, I. C. T. (1970). Autumn migration of the blackpoll warbler: evidence for long flight provided by regional survey. *Bird-Banding*, **41**, 207–40.

Nisbet, I. C. T. & Drury, W. H. (1968). Short-term effects of weather on bird migration: a field study using multivariate statistics. *Animal Behaviour*, **16**, 496–530.

Odum, E. P., Connell, C. E. & Stoddard, H. L. (1961). Flight energy and estimated flight ranges of some migratory birds. *Auk*, **78**, 515–27.

Ogilvie, M. A. (1978). *Wild Geese*. Berkhamsted: T. & A. D. Poyser.

Österlöf, S. (1977). Migration, wintering areas, and site tenacity of the European osprey *Pandion haliaetus*. *Ornis Scandinavica*, **8**, 61–78.

Pennycuick, C. J. (1969). The mechanics of bird migration. *Ibis*, **111**, 525–56.

Pennycuick, C. J. (1972*a*). Soaring behaviour and performance of some East African birds, observed from a motor glider. *Ibis*, **114**, 178–218.

Pennycuick, C. J. (1972*b*). *Animal Flight*. London: Edward Arnold.

Pennycuick, C. J. (1975). Mechanics of flight. In *Avian Biology*, vol. **5**, ed. D. S. Farner & J. R. King, pp. 1–75. New York: Academic Press.

Pennycuick, C. J. (1978). Fifteen testable predictions about bird flight. *Oikos*, **30**, 165–76.

Pennycuick, C. J., Alerstam, T. & Larsson, B. (1979). Soaring migration of the common crane *Grus grus* observed by radar and from an aircraft. *Ornis Scandinavica*, **10**, 241–51.

Perrins, C. M., Harris, M. P. & Britton, C. K. (1973). Survival of Manx shearwaters *Puffinus puffinus*. *Ibis*, **115**, 535–48.

Pienkowski, M. W., Lloyd, C. S. & Minton, C. D. T. (1979). Seasonal and migrational weight changes in dunlins. *Bird Study*, **26**, 134–48.

Porter, R. & Willis, I. (1968). The autumn migration of soaring birds at the Bosphorus. *Ibis*, **110**, 520–36.

Rappole, J. H. & Warner, D. W. (1976). Relationships between behaviour, physiology and weather in avian transients at a migration stopover site. *Oecologia*, **26**, 193–212.

Raveling, D. G. & LeFebre, E. A. (1967). Energy metabolism and theoretical flight range of birds. *Bird-Banding*, **38**, 97–113.

Rayner, J. M. V. (1979). A new approach to animal flight mechanics. *Journal of Experimental Biology*, **80**, 17–54.

Richardson, W. J. (1976). Autumn migration over Puerto Rico and the western Atlantic: a radar study. *Ibis*, **118**, 309–32.

Richardson, W. J. (1978). Timing and amount of bird migration in relation to weather: a review. *Oikos*, **30**, 224–72.

Richardson, W. J. (1979). Southeastward shorebird migration over Nova Scotia and New Brunswick in autumn: a radar study. *Canadian Journal of Zoology*, **57**, 107–24.

Roos, G. (1975). (The migration of arctic waders at Falsterbo in summer 1974 as recorded by three different methods). *Anser*, **14**, 79–92 (Swedish, English summary).

Roos, G. (1977). (Visible bird migration at Falsterbo in autumn 1975). *Anser*, **16**, 169–88 (Swedish, English summary).

Roos, G. (1978a). (Counts of migrating birds and environmental monitoring: long-term changes in the volume of autumn migration at Falsterbo 1942–1977). *Anser*, **17**, 133–8 (Swedish, English summary).

Roos, G. (1978b). (Visible bird migration at Falsterbo in autumn 1976). *Anser*, **17**, 1–22 (Swedish, English summary).

Rudebeck, G. (1950). Studies on bird migration. *Var Fagelvärld*, Suppl. **1**, 1–148.

Salomonsen, F. (1967). *Fuglene pa Grønland*. Copenhagen: Rhodos.

Schnell, G. D. & Hellack, J. J. (1979). Bird flight speeds in nature: optimized or a compromise? *American Naturalist*, **113**, 53–66.

Steidinger, P. (1972). Der Einfluss des Windes auf die Richtung des nächtlichen Vogelzuges. *Ornithologische Beobachter*, **69**, 20–39.

Stewart, A. G. (1978). Swans flying at 8000 meters. *British Birds*, **71**, 459–60.

Summers, R. W. & Waltner, M. (1979). Seasonal variations in the mass of waders in southern Africa, with special reference to migration. *Ostrich*, **50**, 21–37.

Tuck, L. M. (1971). The occurrence of Greenland and European birds in Newfoundland. *Bird-Banding*, **42**, 184–209.

Tucker, V. A. (1973). Bird metabolism during flight: evaluation of a theory. *Journal of Experimental Biology*, **58**, 689–709.

Tucker, V. A. & Schmidt–Koenig, K. (1971). Flight speeds of birds in relation to energetics and wind directions. *Auk*, **88**, 97–107.

Voous, K. H. & Wattel (1963). Distribution and migration of the greater shearwater. *Ardea*, **51**, 143–57.

Ward, P. (1963). Lipid levels in birds preparing to cross the Sahara. *Ibis*, **105**, 109–111.

Welch, A. and L. & Irving, F. (1977). *New Soaring Pilot*, 3rd edn. London: John Murray.

Williams, T. C. & Williams, J. M. (1978). Orientation of trans-Atlantic migrants. In *Animal Migration, Navigation, and Homing,* ed. K. Schmidt–Koenig & W. T. Keeton, pp. 239–51. Berlin: Springer–Verlag.

Williams, T. C., Williams, J. M., Ireland, L. C. & Teal, J. M. (1977). Autumnal bird migration over the western Atlantic Ocean. *American Birds,* **31,** 251–67.

G.P.ARNOLD

Movements of fish in relation to water currents

Introduction

The strategy and tactics of fish migration are discussed in a later chapter. This paper is concerned with logistics. It attempts to review briefly the role of currents in the life histories of migratory fish and to describe some recent discoveries in rather more detail.

Water currents are thought to provide either a transport system or directional clues for migrating fish. Transport can occur by passive drift, if the fish is a pelagic organism with no external reference points, or by modulated drift, if the fish makes vertical movements in the water column. Orientated movements may occur if the fish can detect the speed and direction of the current.

In resolving how fish migrate one fundamental question to be asked is: what are the movements of the fish relative to those of the water at the depth at which the fish is swimming? Some progress has been made during the last decade towards understanding this relationship. One significant advance has been the demonstration that plaice in the North Sea migrate by *selective tidal stream transport*.

Migration and currents

The currents

Four distinct current regimes can be recognized (Fig. 1). Oceanic currents dominate the high seas, while tidal currents are more important on the continental shelves. In estuaries the discharge supplements the ebb tide; in rivers there is only the downstream flow.

The surface circulation of the main ocean basins is driven largely by the prevailing winds. Its basic pattern is a nearly closed system called a gyre. The northern and southern limits of each gyre are marked by zonal currents flowing east and west. These are joined by meridional boundary currents setting nearly parallel to the continental margins. In the northern hemisphere the Atlantic and Pacific oceans have large subtropical and somewhat smaller subpolar gyres centred at approximately 30° and 50° N. Open ocean currents generally flow at speeds of 3–6 km/day (3.5–6.9 cm s^{-1}) and normally extend to depths of 100–

200 m. But the western boundary currents, which flow northwards in the northern hemisphere, are very much stronger. The Gulf Stream and Kuroshio currents in the North Atlantic and North Pacific Oceans, respectively, are rather narrow jet-like currents flowing at speeds of 40 and 120 km/day (46 and 139 cm s^{-1}), and extending to depths of 1000 m or more. Substantial counter-currents occur below the surface circulation in many places.

Tidal currents, which are generated by the combined gravitational effects of sun and moon, are present in both oceanic and shelf waters. But their effects are much more pronounced on the continental shelves. Near coasts the rotatory tides of the open sea are replaced by the familiar reversing tides of the seaside. *Flood* tides, where the water flows towards the coast, are separated by periods of slack water from *ebb* tides, where water flows away from the coast. Three types of tide can be recognized: *diurnal* tides with one high and one low water per tidal day of 24 h 50 min; *semi-diurnal* tides with two high and two low waters per tidal day and a tidal period of 12 h 25 min; and *mixed* tides with appreciable differences in height between successive high and low tides and unequal tidal periods. Semi-diurnal tides are typical of the Atlantic coasts of Europe and North America, mixed tides of the Pacific coast of North America. The daily tidal range is greatest during *spring tides,* which occur at times of new and full moon, and least during *neap tides,* which occur near the first and third lunar quarters. The speeds of the tidal currents vary correspondingly.

An estuary is a semi-enclosed tidal basin, in which the circulation is a mixing system for fresh and saltwater. In a simple *salt-wedge* estuary, where the river flow is dominant, there is a vertical stratification. A wedge of seawater extending upstream on the bottom is separated by a sharp discontinuity from freshwater flowing seawards above. Where the river flow is less dominant there is still a two-layered circulation but entrainment and vertical mixing break down the interface and produce a salinity gradient which may be localized in the *halocline* or which may extend throughout the water column. Where the tidal current is dominant the intensity of vertical mixing is such that both the salinity gradient

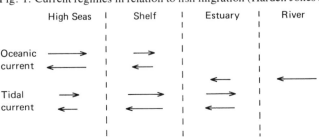

Fig. 1. Current regimes in relation to fish migration (Harden Jones *et al.,* 1978).

and the vertical stratification break down. The seaward and landward net flows are then separated horizontally rather than vertically. Seasonal fluctuations in river discharge vary the degree of tidal influence.

The migration pattern

In general terms the movements of migratory fish are related to those of the water currents. Life histories are geared to regional production cycles and stocks are contained within oceanic gyres (Harden Jones, 1980). In temperate and arctic waters there are discrete spawning, nursery and adult feeding grounds and, concomitantly, distinct spawning, feeding and wintering migrations. Eggs and larvae drift downstream with the residual current to the nursery grounds and at a later stage in the life history the adult fish must make a compensatory movement if the population is to remain within the same hydrographic system. These movements are described as *denatant* and *contranatant,* respectively (Meek, 1915). Denatant means swimming, drifting or migrating with the current; contranatant means swimming or migrating against the current. In the classic contranatant theory of fish migration (Russell, 1937) it is inferred that prespawning migrations are active contranatant movements against the prevailing current, while postspawning migrations of spent fish are passive denatant movements with the prevailing current.

Life histories

Diadromous fish such as salmon and eels migrate between the sea and freshwater and pass through all four current regimes during their life history. Many more species of truly migratory fish are limited either to the sea or to freshwater. Life histories vary accordingly and some typical examples follow; others are described in this book in the chapter by Harden Jones.

Most species of salmon are *anadromous* spending much of their life in the sea and only returning to freshwater to spawn. The eggs are laid in gravel beds, or redds, and in most species the young fish migrate to sea as *smolts* after two or three years in nursery streams or lakes. In two species of Pacific salmon the young migrate to sea as *fry* shortly after hatching. Eels are *catadromous,* spawning in the sea and spending most of their adult lives in freshwater. The European eel, *Anguilla anguilla,* spawns in the Sargasso Sea and experiences an extensive drift migration across the North Atlantic during its larval stages. Newly metamorphosed elvers arrive in European coastal waters two and a half years after spawning.

Among pelagic fish there are five species of tuna – bluefin, albacore, bigeye, yellowfin and skipjack – which occur more or less continuously across the major oceans and which make extensive migrations. In the North Pacific the albacore, *Thunnus alalunga,* spawns in the central subtropical zone. It performs transpacific migrations and supports important Japanese and American fisheries. There is probably one stock contained within the subtropical gyre. On both sides of the Pacific Ocean the distribution of albacore appears to be related to the fronts between water masses: in the north they feed in the areas of divergence towards the edge of the subtropical gyre and in the south in the areas of convergence in the equatorial region.

The plaice, *Pleuronectes platessa*, which is found on the sandy parts of the continental shelf of Europe, is a typical demersal flatfish, whose life is greatly influenced by both residual and tidal currents. It spawns in winter producing large numbers of small pelagic eggs. The larvae reach metamorphosis at the end of the planktonic phase. They then become asymmetrical and drop to the bottom as miniature flatfish before moving inshore onto intertidal nursery grounds. Tidal transport is a recurrent feature of the plaice's life history from metamorphosis onwards. The migrations of southern North Sea plaice are described in more detail below.

Reactions to the current

Flatfish lying on the seabed can detect the current as a result of direct mechanical stimulation. But a fish not in contact with the bottom requires a visual reference point and must be within a few metres of the river bank or seabed to detect its displacement. Fish can detect current speeds of $1-10$ cm s^{-1} (Arnold, 1974).

There is no evidence that fish out of sight of the bottom can detect the current – Westerberg's (1979) observations are inconclusive – but fish in midwater in the sea might be able to orientate visually at the *rheocline* (the velocity gradient occurring vertically across the thermocline) if particles are visible in the water (Harden Jones, 1968). Alternatively they might be able to detect the electric fields generated by the movement of water through the vertical component of the earth's magnetic field. Skates and rays and salmon and eels can detect potential gradients of $0.01-0.1$ μV cm^{-1}; gradients of $0.05-0.5$ μV cm^{-1} commonly occur in the ocean (Kalmijn, 1971; Rommel & McCleave, 1973a, b).

The basic reaction of fish to currents is *rheotaxis*. Most fish turn to head into the current and maintain station relative to a fixed reference point. As the speed of the current increases the fish swims faster; a kinesis is superimposed on the taxis. But this behaviour must be modified for migration and at different stages of their life history diadromous fish move both upstream and downstream. Loss of visual reference points at night may account for the downstream displacement of salmonid fry. But the seaward movements of smolts and adult eels are apparently active orientated movements; the fish head downstream and swim with the current.

There is some evidence that diadromous fish show behavioural changes at the interfaces between the current regimes. A delay at the freshwater/saltwater boundary is common in both adult and juvenile salmonids. Silver eels change from negative to positive rheotaxis on exposure to saltwater (Hain, 1975). Newly metamorphosed elvers at first drift passively with the tide in coastal waters, progressing only with the residual current. Subsequently, they switch successively to tidal transport and active upstream movement (Deelder, 1958). The second change occurs, in the River Elbe at least, at the limit of the tidal incursion (Tesch, 1965).

Oceanic migrations are generally consistent with movements with, rather than against, the current. Because they are not well described in relation to oceanographic features, they can be accounted for by a number of different models. These variously invoke passive drift, active downstream migration or some form of independent navigation. Passive drift in the North Equatorial Current and Counter Current can, for example, account for the movements of young skipjack tuna, *Katsuwonus pelamis,* in the Pacific Ocean and jet-stream transport can explain those speeds which are significantly faster than average (Seckel, 1972; Williams, 1972). Two bluefin tuna, *Thunnus thynnus,* which travelled at least 4832 km between the Bahamas and Bergen in Norway at an average speed of 40 km/day (Mather, 1962), probably received substantial assistance from the Gulf Stream and North Atlantic Drift. But in several instances ground speeds estimated from conventional tagging experiments suggest that oceanic migrations cannot be accounted for by drift alone. Sockeye salmon, *Oncorhynchus nerka,* in the eastern North Pacific Ocean travel at speeds ranging from 20 to 50 km/day (French, Bilton, Osako & Hartt, 1976). Two albacore covered 5239 km from the central North Pacific to the coast of Washington State in 110 days at an average speed of 48 km/day (Sharp & Dotson, 1977). Because surface currents in these areas have much lower average speeds it is argued that these are active orientated movements. Royce, Smith & Hartt (1968) suggest that salmon follow the ocean circulation in the North Pacific but swim downstream faster than the current, which they detect by means of electric potentials. Neave (1964) suggests that some form of bicoordinate navigation may be required.

Selective tidal stream transport in plaice
The Southern Bight stock

In the North Sea there are four stocks of plaice which spawn off the Scottish east coast, Flamborough Head and in the Southern and German Bights respectively. There are also spawnings in the English Channel. In the summer the Southern Bight, German Bight and Flamborough spawners occupy different feeding areas with only a limited overlap to the south of the Dogger Bank.

The Southern Bight stock spawns in winter and in November and December mature plaice move some 200–300 km south from the Leman Ground and more northerly areas to a spawning area centred on the Hinder Ground (Fig. 2) and bounded by latitudes 51°30′–52°00′ N and longitudes 2°–3° E. Peak egg production occurs in January and the spent fish return north in January, February and March. Meanwhile the pelagic eggs and larvae drift to the northeast at a rate of about 3 km/day with the residual current entering the North Sea through the Straits of Dover. Depending on temperature the larvae take about 5–8 weeks to reach metamorphosis. The position at which they drop to the bottom is governed by the speed and direction of the residual flow, which under conditions of strong and persistent northeasterly winds (8–10 m s⁻¹) may be checked or even reversed. Most of the larvae take to the bottom along the Dutch coast, frequently some distance offshore, but in some years close to Texel (Harding, Nichols & Tungate, 1978), where under certain conditions there may be an onshore residual current at the bottom (Cushing, 1972).

Many of these young plaice enter the Dutch Wadden Sea through the inlets between the islands of Texel, Vlieland and Terschelling. The Wadden Sea, which extends from Den Helder in Holland to Esbjerg in Denmark and covers

Fig. 2. Place names in the southern North Sea of interest in connection with plaice migrations. The positions of tidal stations N and P (Fig. 8) are indicated by diamonds. The 30-m depth contour is indicated.

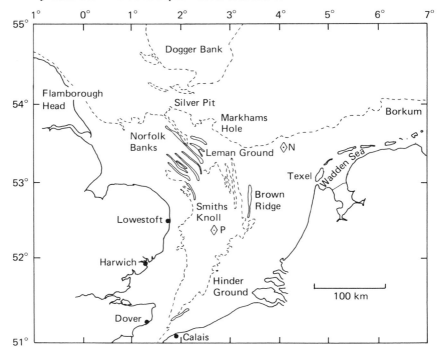

an area of some 8000 km², is an area of deep creeks and extensive tidal mudflats with a very rich bottom fauna. Other important nursery areas are the estuaries of the Maas and Scheldt and the sandy beaches of the open coast.

From summer depths of 2–3 m the plaice move to deeper water in autumn returning to shallower water the following spring and early summer. There is an overall movement to deeper water in their second and third years of life. Some of the larger plaice, mostly males, reach first maturity in their third year of life, joining the mature fish of earlier year-classes on the spawning grounds towards the end of the season. These first-time spawners migrate down the eastern side of the Southern Bight between Texel and the Brown Ridge (Fig. 2). After spawning, the first-time and repeat spawners migrate north to the feeding grounds moving up the western side of the Southern Bight. Most of the remaining immature fish spawn for the first time the following year and recruitment to the adult stock is then almost complete. Plaice spawn annually until they die but with present levels of exploitation an individual would be unlikely to spawn more than ten times.

Movements of adult fish

Adult plaice fitted with 300-kHz transponding acoustic tags (Mitson & Storeton-West, 1971) and tracked in the open sea by sector-scanning sonar (Voglis & Cook, 1966; Mitson & Cook, 1971) show a pattern of semi-diurnal vertical movement related to the tides, which we have called *selective tidal stream transport* (Greer Walker, Harden Jones & Arnold, 1978).

The fish were tracked within an area bounded by latitudes 51°40' N and 52°40' N, longitudes 1°40' E and 2°30' E (Fig. 2). The tides in this area are fast, with speeds up to 1.4 m s⁻¹ at springs, and highly directional, flowing approximately northnortheast and southsouthwest. The tide floods on the south-going stream and the tidal range is 3 m in water of 20–40 m depth. The bottom deposits are mainly fine to medium sand with beds of gravel in some places and patches of sand waves (wavelength 2–3 m) in others. The water is generally turbid and visibility is limited to a few metres.

The fish were caught by otter trawl in the Silver Pits and on the Borkum Ground (Fig. 2) and were held in the laboratory for about two months prior to release. Twelve fish (mean length 41.3 cm; s.d. 4.6 cm) were released singly and followed for periods of 25–55 h and distances of 9–61 km. Eight fish which covered distances greater than 15 km showed a consistent pattern of tidal transport. They made regular vertical migrations, coming off the bottom at one slackwater and remaining in midwater for the following 5–6 h before returning to the bottom at the next slack water. When in midwater the fish moved downstream with the tide for distances of 12–16 km, more or less equal to that of the tidal displacement. Activity was limited when the fish were on the bottom and some fish did not move for 2–3 h. The regular cycle of movements on and off the bottom at slack water was very striking and this pattern of semi-diurnal vertical migration was characteristic of all fish that moved appreciable distances.

The track of plaice no. 7 was typical (Figs. 3 and 4). The fish came up into midwater after high water slack, gained ground to the north during the north-going tide and returned to the bottom at low water slack. The speed of the fish over the ground was 80, 68 and 105 cm s^{-1} for the three northerly tides compared with only 8 and 9 cm s^{-1} for the two south-going tides.

Fig. 3. Track chart of plaice no. 7, released at 1009 h, 12 December 1971. Hourly positions of the fish are indicated and the times of slackwater are given. Open circle, north-going tide; circle with upper segment filled, low water slack; filled circle, south-going tide; circle with lower segment filled, high water slack. The 10 m and 20 m depth contours are shown. (Simplified from Greer Walker *et al.*, 1978.)

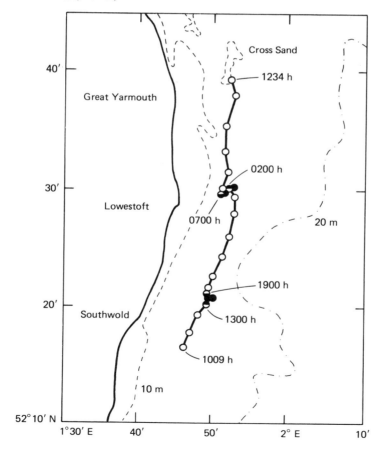

During this, and some other tracks, water-current measurements were made from a second research ship or with moored current meters. From these measurements it was possible to estimate the speed of the fish through the water. It rarely exceeded one body length per second (1 L s^{-1}) and, while fish no. 7 appeared to swim downstream with the tide, three other fish swam through the water heading consistently southeast as they were being carried northwards over the ground. The orientation was maintained for periods up to one hour and may well have been determined with reference to sandwaves before the fish left the bottom. More recent observations, where the heading of the fish is returned to the research ship by telemetry, are discussed by Dr Harden Jones in his chapter of this book.

The spawning migrations

If plaice used selective tidal stream transport when on migration, ripening fish moving towards their spawning grounds in the Southern Bight during November and December should be in midwater on south-going tides, whereas spent fish returning to their feeding areas in February and March should be in midwater on north-going tides. This prediction has been tested by midwater trawling experiments along the line of the western migration route (Harden Jones, Arnold, Greer Walker & Scholes, 1979).

The experiments were carried out with large midwater trawls. Paired tows, usually of 3 h duration, were made on successive tides and the gear was shot so as to start fishing at the required depth one and a half hours before the expected time of maximum tidal speed. Each haul was made downtide along the tidal axis and, so far as was possible, the research vessel covered the same track over the ground on each tow of a pair.

Fig. 4. The depth of plaice no. 7 in relation to the direction of the tide and other environmental factors. (Simplified from Greer Walker *et al.*, 1978.)

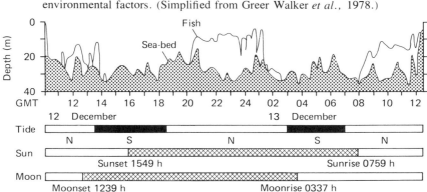

The work was carried out on the western side of the Southern Bight and most plaice were caught in the line of the 'Knoll Stream' between latitudes 52° N and 52°40' N, longitudes 2°00' E and 2°30' E. At the centre of this area the directions of the north-going and south-going tides are 020° T and 200° T, respectively; peak speeds of 70 and 120 cm s^{-1} are reached at neap and spring tides.

Fifty paired tows were made in November and December ('autumn' pairs) and 22 in February ('winter' pairs) giving 39 and 21 pairs, respectively, in which plaice were caught in one or both tows. More plaice were caught by night than by day but this diurnal effect was removed by restricting the comparison to the 17 'autumn' and 8 'winter' pairs in which both tows were completed during the night. There were no pairs in which both tows were completed during the day. During the prespawning migration significantly more fish were caught on the south-going tide; during the postspawning migration significantly more fish were caught on the north-going tide (Table 1).

In the 'autumn' most of the female plaice were still maturing and none was yet ripe, whilst half the males were already 'running ripe' and extruding sperm. No spent fish were caught in 'autumn' but three-quarters of the females caught in February had completed spawning. Virtually none of these fish had been feeding.

These midwater trawling experiments provide direct evidence that in November and December ripening plaice migrating into the Southern Bight by the western route use tidal stream transport from at least as far north as the Leman Ground. The paired tows made in February show that spent plaice return by the same transport system, at least as far as Smiths Knoll.

Diurnal vertical movements on spawning and feeding grounds
Plaice kept in aquaria under natural illumination show a clear rhythm of diurnal vertical migration, feeding on the bottom during the day and swimming

Table 1. *Catches of plaice by night in a midwater trawl on consecutive north-going (N) and south-going (S) tides in the southern North Sea from November 1974 to February 1977 (Harden Jones et al., 1979)*

Months	No. of paired tows	No. of plaice caught on each tide		Catch rate on each tide		Paired t-test		
		N	S	N	S	t	d.f.	P
Nov, Dec	18	35	149	1.9	8.3	4.42	17	<0.001
Feb	8	84	11	10.5	1.4	4.36	7	<0.01

freely off the bottom at night (Verheijen & de Groot, 1967). During most of the year catches of plaice in bottom trawls in the Southern Bight are lower by night than by day with an average day:night catch ratio of 1.3:1.0. But on the spawning grounds, in January and February, this ratio is reversed (de Groot, 1971). Plaice are quite commonly caught well above the bottom in midwater trawls and it is generally assumed that the reduction in catch at night is partly or mainly the result of a diurnal vertical migration, whereby at night a proportion of the population is no longer accessible to gear on the seabed. The reversal of the catch ratio during the spawning season may perhaps be attributed to an alteration or even a reversal of this behaviour.

Recent experiments (G. P. Arnold, M. Greer Walker & F. R. Harden Jones, unpublished results) have confirmed a diurnal pattern of vertical migration in plaice on their feeding grounds in the vicinity of Markhams Hole (Fig. 2). For example, one fish (plaice no. 2, RV CLIONE cruise 6/79), which was tracked for 82 h (Figs. 5 and 6), made four vertical excursions into midwater. It left the bottom each night at, or shortly after, sunset and swam freely in the water column for periods of 1–3 h. Apart from an initial period of eight hours after release, movement was restricted to the periods when the fish was in midwater. It covered a total distance of 26.5 km at an average ground speed of 0.3 km h^{-1}. This was five times less than that of plaice no. 7, whose track has been described above, and which covered an overall distance of 43 km in 26 h by tidal transport. The tides in the Markhams Hole area run northwest and southeast and have peak speeds of 35 and 57 cm s^{-1} at neaps and springs respectively. These speeds are about half those encountered off the East Anglian coast.

Movements of juvenile fish

Selective tidal stream transport is probably an innate pattern of behaviour in the plaice, occurring as it does with other rheotactic behaviour at metamorphosis, immediately after the larvae have gone to the seabed. It has been demonstrated for larvae entering the Wadden Sea in March and April through the Marsdiep inlet between Texel and Den Helder. Catches of larvae in an Isaacs–Kidd midwater trawl were consistently higher on the flood than on the ebb tide and the catch ratio increased progressively with the age of the fish. Less than 20% of fully metamorphosed Stage 5 larvae were found in midwater on the ebb tide compared with over 65% of the early Stage 4 larvae, which are only just beginning to metamorphose (Creutzberg, Eltink & van Noort, 1978).

After spending their first summer in the Wadden Sea the O-group plaice move out to sea again in the autumn to overwinter offshore. Some return again as I-group fish in the following February and March. These fish also use selective tidal stream transport, moving landwards through the Marsdiep and Vliestroom inlets on the flood tide (de Veen, 1978).

As both O- and I-group fish the plaice in the Wadden Sea show a tidal feeding migration, collecting in the creeks at low tide but moving up onto the mudflats as they are covered by the rising tide (Kuipers, 1973). Similar tidal migrations occur on sandy beaches (Gibson, 1973).

Behavioural problems

The extensive use by the plaice throughout its life history of tidal currents for transport raises a number of behavioural problems. For local movements within a feeding area, direction is probably unimportant and a diurnal vertical migration, occurring independently of the tide, will probably achieve adequate dispersal. But direction – and thus choice of tide – must be of paramount importance when plaice enter or leave a nursery area, or when they

Fig. 5. Track chart of plaice no. 2 (RV CLIONE cruise 6/79) released at 1506 h, 27 May 1979. During the day the fish remained within the area indicated by the large open circles. Hourly positions of the fish when it was moving are indicated by the smaller open (day) and closed (night) circles. The portion of the track when the fish was moving on the bottom is indicated by the heavy line. The 30-m depth contour is shown.

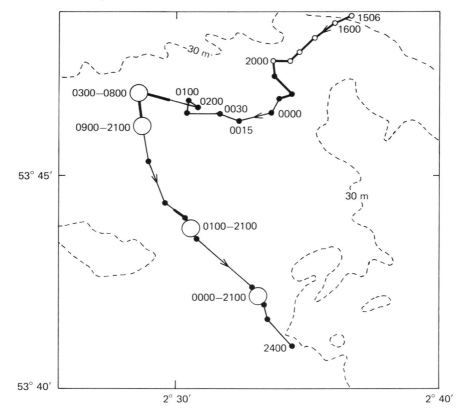

migrate between feeding and spawning grounds. The fish must then be able at the outset to 'lock-on' to the appropriate tide.

Newly metamorphosed fish show a diurnal rhythm of activity until they enter the intertidal zone (Gibson, Blaxter & de Groot, 1978). It is only then that a semi-diurnal tidal activity cycle appears and this rapidly reverts spontaneously to the basic diurnal, or circadian, rhythm in the absence of appropriate entraining stimuli (Gibson, 1973, 1976). A fish joining the tidal transport system must switch from a diurnal to a semi-diurnal pattern of vertical migration and it seems that this switching ability must be a basic feature of the plaice's life history. Hydrostatic pressure cycles are at least partially responsible for entraining circatidal cycles of activity in the intertidal zone (Gibson *et al.*, 1978) but it is not known what the entraining stimuli are in the open sea. Clearly, though, if the fish is to move in the direction appropriate to its migration, the entraining signal must be unambiguous and there must be a distinction between high and low water slack tides. The south-going tide which, in the Southern Bight, transports the ripening fish south to their spawning grounds and the recently metamorphosed fish onto their nursery grounds, is preceded by low water slack. The north-going tide, which returns the spent fish to their feeding grounds, is preceded by high water slack. One possible entraining stimulus for the ripening fish is the conjunction of sunset, full moon and low water slack, which occurs for 2–3 days every month (Harden Jones, 1979). Cessation of feeding might also be a contributory factor. Starvation appears to be an essential precursor of tidal transport in recently metamorphosed larvae. It is not known how these larvae distinguish between ebb and flood tides but, unlike elvers and crabs (p. 72), it is apparently not in response to changes in odour or salinity (Creutzberg *et al.*, 1978).

Fig. 6. The depth of plaice no. 2 (RV CLIONE cruise 6/79) in relation to the direction of the tide and the time of sunset. The solid bars indicate a tide flowing to the southeast; the open bars a tide flowing to the northwest.

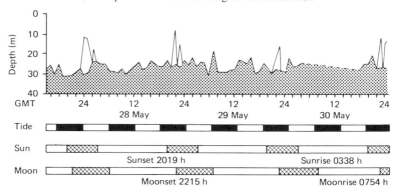

Some behavioural mechanism must maintain the semi-diurnal pattern of vertical movement so that the fish can repeatedly join and leave the appropriate tide during the course of its migration. Ascents are more closely related to slack water than are descents, and descents are often preceded by excursions to the bottom (Greer Walker *et al.*, 1978). These observations suggest that a rheotactic response mediated through sight of, or contact with, the bottom and a semi-diurnal rhythm of activity entrained to a tidal cycle are involved in the synchronizing mechanism.

Plaice must leave the transport system on arrival at their destination. In the case of feeding and nursery grounds location may not be particularly important and the inhibition of pelagic swimming with the onset of feeding, as is observed in metamorphosing larvae, might be sufficient to terminate the movement. Spawning grounds may, however, be more limited and require the recognition of local landmarks. Chemical clues are involved in the return of salmonids to their home rivers and, following the homestream odour theory, Harden Jones (1979) has suggested that the assembly and spawning areas of marine fish in coastal waters could be identified by chemical clues entering the sea by groundwater seepage. The hypothesis is discussed in the chapter by Harden Jones in this book.

The generality of the tidal stream transport hypothesis

Selective tidal stream transport occurs in other species as well as plaice and is probably of general significance in shelf seas, wherever tidal currents are reasonably fast and directional.

A tidal streampath chart

Although tidal stream transport has been demonstrated in plaice of Borkum origin released and tracked off the East Anglian coast (Greer Walker *et al.*, 1978), further midwater trawling experiments are needed to demonstrate that the mechanism is used by the plaice populations that spawn off Flamborough and in the German Bight. Theoretically, though, it seems possible that the stocks of plaice in the southern North Sea might be contained along the lines of the tidal streampaths (Harden Jones, Greer Walker & Arnold, 1978), rather than within the boundaries of a residual current system. The tidal streampaths of the North Sea (Fig. 7) clearly separate the Southern and German Bight stocks and link the Flamborough plaice with those of the Southern Bight and the English Channel. The streampaths also suggest that if a silver eel (p. 71) were to adopt selective tidal transport off the east coast of England it could readily be transported through the Channel and out into the Atlantic Ocean through the Western Approaches.

The directivity of the tidal streams is also relevant to the tidal transport hypothesis. Directivity can be measured by expressing, at one station, the slowest tidal speed as a percentage of the fastest. Thus long thin tidal ellipses, where the speed of the slowest tide is less than 10% of the fastest, are directional; rounded ellipses, where the slowest tide may be as much as 60% of the fastest, are less so. In the Southern Bight Admiralty tidal stations N and P (Fig. 2) are representative of areas of low and high directivity respectively (Fig. 8(a)) and progressive vector diagrams (Fig. 8(b)) show that fish using selective tidal stream transport will travel much faster in one area than the other. It may thus be more than coincidence that the western migration route of plaice moving to and from the Southern Bight spawning grounds goes through the area off the East Anglian coast, where the tidal ellipses are most directional.

Fig. 7. A tidal streampath chart for the North Sea and adjacent areas (Harden Jones *et al.*, 1978). The centres of the Flamborough, German Bight, Southern Bight and eastern English Channel plaice spawnings are indicated by the black circles.

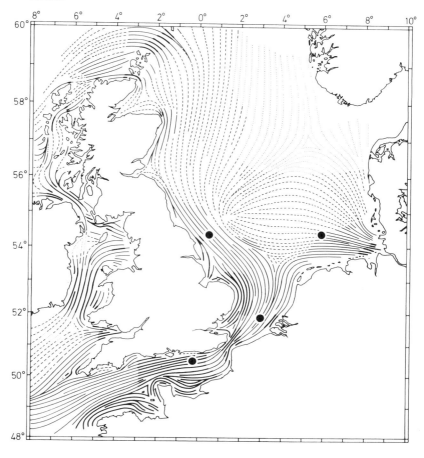

Fig. 8(a). Ellipses at Admiralty tidal stations N (53°27.2' N 04°05.1' E) and P (52°24.0' N 02°41.8' E) in the Southern Bight of the North Sea (see Fig. 2). The speed and direction of the tide is shown as a vector for each hour before ($-$) and after ($+$) High Water Dover (HWD). Speed is given for spring tides. (Data from International Chart 2182 A.) (b). Progressive vector diagrams for fish using selective tidal stream transport at tidal stations N and P. A fish using tidal transport would travel considerably faster at station P than at station N.

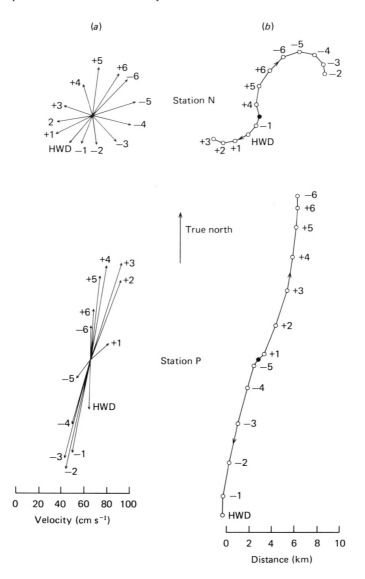

It may also be that there are areas where the physical characteristics associated with rounded ellipses – lower velocities, or smaller linear and angular accelerations – are not strong enough to elicit tidal transport. Acoustically tagged plaice released in the summer on grounds to the northwest of Texel, where the index of directivity of the tidal ellipses is low, did not show this behaviour (Harden Jones, Margetts, Greer Walker & Arnold, 1977).

The speeds at which fish travel on migration are generally rather slow (Harden Jones *et al.*, 1978) and this is certainly true for the plaice populations in the southern North Sea (de Veen, 1978). Conventional tagging experiments suggest that the bulk of a population may take several months to cover a distance that could be achieved in 1–2 weeks by sustained tidal transport. It seems likely, therefore, that the mechanism affords plaice a means of making rapid directed movements at selected times and places and that, at any one time, only a fraction of the population may be in transit.

Tidal stream transport in other species

Adult prespawning soles (*Solea solea*) are sometimes observed in the southern North Sea drifting passively at the surface at night when the tide is flowing towards their spawning grounds along the Dutch, German and Danish coasts. de Veen (1967) suggests that these fish leave the bottom only at night. But it may well be that this migration also occurs by tidal stream transport. The mechanism has recently been demonstrated in sole (Greer Walker, Riley & Emerson, 1980). Other recent unpublished work at the Fisheries Laboratory, Lowestoft, has demonstrated selective tidal stream transport in cod, *Gadus morhua*, during their postspawning migration in the Southern Bight and in silver eels, *Anguilla anguilla*, released off the East Anglian coast.

The tracks of some shad, *Alosa sapidissima*, returning to the Connecticut River from Long Island Sound (Dodson & Leggett, 1973) are consistent with tidal stream transport as are those of some eels in the German Bight (Tesch, 1974) and some Pacific salmon in the coastal waters of British Columbia (Stasko, Horrall, Hasler & Stasko, 1973). But in all cases the essential depth information is lacking. Tidal stream transport occurs in red seabream, *Chrysophrys major*, passing through the Naruto Strait between the islands of Awaji and Shikoku in Japan (Ichihara, 1979).

Energy savings

It can be shown theoretically (Weihs, 1978) that in areas where tides are directional fish may save up to 40% of the energy cost of the journey by using tidal stream transport compared with the alternative strategy of swimming continuously in the direction of overall movement. For a 1-kg female plaice covering a distance of about 280 km between its feeding area south of the Dogger

Bank and the spawning area in the Southern Bight the total energy cost of the prespawning migration is equivalent to 6 g of fat. A saving of 2.4 g of fat is equivalent to 1% of the weight of the gonad and represents a potential increase in fecundity, or higher reserves for some or all of the eggs, which might be expected to have a selective advantage over a few hundred generations (Harden Jones, 1980).

Tidal stream transport in estuaries

The tidal migrations of juvenile plaice in and out of the Wadden Sea have been described above. Similar seasonal migrations are made in the same area by other flatfish and also a number of invertebrates including the swimming crab *Macropipus holsatus* (Verwey, 1958). The movements of O- and I-group flounders, *Platichthys flesus*, and soles follow the same pattern as those of plaice; those of dab, *Limanda limanda*, are a little different (de Veen, 1978).

Each spring, too, large numbers of elvers enter the Wadden Sea through the Marsdiep and migrate inshore to places where there are sluices discharging freshwater. During the flood tide the elvers are off the bottom and are transported passively inshore, but on the ebb tide they are on the bottom and are not displaced seaward again (Creutzberg, 1961). Laboratory experiments have shown that the elvers discriminate between the ebb and flood streams by an olfactory stimulus from freshwater, which is present only during the ebb tide. The crab, *M. holsatus*, in contrast, distinguishes between ebb and flood tides by changes in salinity (Venema & Creutzberg, 1973), as does the pink shrimp *Penaeus duorarum*, which makes similar tidal migrations in Florida (Hughes, 1969).

Larval herring, *Clupea harengus*, move up the Sheepscot estuary in Maine in spring and autumn, accumulating in the upper portion of the lower estuary, where they are retained despite the residual seaward flow. This estuary has a two-layer flow with vertical mixing. The greatest net excursion is near the surface on the ebb tide and below mid-depth on the flood tide; an area of no net excursion is located at mid-depth (Graham & Davis, 1971). It is inferred from catches of buoyed and anchored plankton nets that the larvae initially enter the estuary in the net landward flow near the bottom. Subsequently they move vertically in the water column and are transported seawards on the ebb tide at the surface and landwards on the flood tide near the bottom (Graham, 1972).

The movements of adult anadromous fish returning to their home rivers also appear to be related to the tide but here the situation is complicated by a period of estuarine delay, during which the fish adapt physiologically to freshwater. Atlantic salmon, *Salmo salar*, in the partially mixed estuary of the Miramichi, New Brunswick (Stasko, 1975), sockeye salmon in the salt-wedge estuary of the Skeena River in British Columbia (Groot *et al.*, 1975) and shad in the salt-wedge estuary of the Connecticut River (Leggett, 1976) drift passively to and fro with the tide in the vicinity of the saltwater–freshwater interface. But some fish swim against both ebb and flood tides and those which make progress upstream into freshwater do so by holding ground on the ebb tide and drifting or swimming upstream with the flood. Such movements occur in the deeper estuarine channels and salmon are said to hug the bottom following the deepest channel – the *thalweg* – where the current speed is usually least (Harden Jones, 1980). Once into freshwater, shad swim much faster than in brackish water and appear to vary their speed in response to tide-induced variations in current velocity. This results in an approximately constant ground speed and suggests that the movement is regulated by a typical visual or optomotor response. Energetically, such behaviour may be less efficient than the alternative strategy of maintaining a constant speed through the water (Leggett & Trump, 1978). Similar behaviour is thought to occur in the waters of Long Island Sound before the fish reach the estuary of the Connecticut River but after they have come within the influence of its water. Olfactory stimuli are implicated in this behaviour (Dodson & Leggett, 1974) and also in that of the alewife, *Alosa pseudoharengus*. This clupeid distinguishes its homestream water from that of other streams by olfaction (Thunberg, 1971) suggesting that Hasler's (1960) homestream odour hypothesis is applicable to anadromous fish generally and not just to the salmonids.

Direct observations (White & Huntsman, 1938) and netting experiments (Calderwood, 1905; Berry, 1932) suggest that Atlantic salmon smolts select the ebb tide on which to initiate their seaward movement through the estuary. But the movement itself appears to be passive (Tytler, Thorpe & Shearer, 1978). In the upper, riverine, portion of the estuary of the Penobscot River in Maine, hatchery-reared smolts moved to and fro with both ebb and flood tides. The fish were near the surface and their net rate of seaward movement was very close to that of drogues drifting at the same depth (Fried, McCleave & LaBar, 1978). Similar movements occurred in the upper part of Penobscot Bay (LaBar, McCleave & Fried, 1978). But in the lower part of the bay some seaward movement occurred during the flood tide and some fish appeared to progress faster than the ebb tide, which suggests a change to an active seaward movement. Most fish moved in the main current away from the shoreline and covered 45–57 km from head of tide (salinity $0^0/oo$) to open water (salinity $29^0/oo$) in less than 48 h. Some juvenile Pacific salmon make tidal feeding migrations in estuarine conditions (Mason, 1974) but a period of physiological adaptation in brackish water does not seem to be necessary at this stage of the life history.

The ability to respond differently to ebb and flood tides in estuaries and adjacent coastal waters must be a key factor in the lives of diadromous and other fish which make estuarine migrations. It is obviously not fully understood how such movements are related to the various types of estuarine circulation and further tracking experiments, in which the depth of the fish can be determined, are clearly desirable.

Other mechanisms of migration

Selective tidal stream transport is probably the optimal strategy for fish migrating in tidal waters. It clearly allows rapid directional movements with a minimal expenditure of energy, very probably without the need for any navigational ability. But it is not the whole story even on the continental shelf and it is unlikely to be applicable in the open ocean.

On the shelf both plaice and cod show contranatant movements, swimming on some occasions against the adverse tide between periods of midwater transport. Such movements invariably occur while the fish is on or close to the bottom and has the necessary visual or tactile stimulus with which to detect the current. But fish out of sight of the bottom may nevertheless swim consistently in one direction: plaice can do so for short periods, salmon for several tidal cycles. Adult pink, *O. gorbuscha* (Stasko *et al.*, 1973), and sockeye (Stasko, Horrall & Hasler, 1976) salmon returning to the Fraser River in British Columbia migrate actively through the confined waters of the San Juan archipelago along the axes of the tidal currents. Recent unpublished work of the Fisheries Laboratory, Lowestoft, has shown that Atlantic salmon behave similarly. Two grilse tracked off the northeast coast of England in 1979 moved consistently northwards, swimming mostly within four metres of the surface. The speed of the fish over the ground was very much greater on the north-going tide but the fish continued to make progress against the south-going tide with their courses virtually unchanged. Plaice may obtain their orientation before leaving the bottom either directly from the current itself or secondarily from some current-generated feature of the bottom topography. But how the salmon orientate is unknown and the current may or may not be implicated.

Few fish tracks are as yet available from the oceanic regime. Japanese work has shown that yellowtails, *Seriola quinqueradiata,* while swimming to depths of 180 m and traversing the thermocline with ease, nevertheless move downstream with the current (Ichihara, 1979). A solitary chum salmon, *O. keta,* tracked off the southern Kurile Islands for a period of 37 h in the autumn of 1974 behaved similarly (Ichihara, Yonemori & Asai, 1975). It made frequent vertical movements in the top 45 m of the water column (depth 100 m) and followed a convoluted path. The total length of its track was 164 km compared with a point-to-point distance of only 35 km and it swam at a speed of $1.8-1.9\,L$ s^{-1} (L, 68 cm). It moved to the southwest off Etorofu Island and was displaced with the current that flows in this direction along the southern side of the Kurile Islands in the autumn. At this time of year chum salmon are returning to Hokkaido and

Honshu in the Kurile Current. The tracks of five bluefin tuna followed offshore from Halifax, Nova Scotia, by Lawson & Carey (1972) were rather different. They were straight for many hours on end and movement was regular by day and night. But the tracks radiated in a number of different directions, suggesting that while the fish could maintain a consistent direction they were not orientated to any common celestial or hydrographic feature. One fish, however, made frequent excursions through the thermocline; this suggests a similarity of behaviour with Pacific albacore tracked in US coastal waters (Laurs, Yuen & Johnson, 1977). These fish showed a tendency to collect in the vicinity of upwelling coastal fronts, presumably to feed, and moved away from the immediate area when upwelling ceased. Their movements also appeared to be related to the distribution of surface temperature; they spent little time in water where this was less than 15°C.

Two parallels can be drawn for these oceanic migrations, one with the migrations of *Schistocerca gregaria,* the desert locust (Rainey, 1978), and the other with the ocean voyages of the early European navigators (Heyerdahl, 1978). The former is a pure drift migration; the latter a combination of drift and compass course.

The Asiatic coasts of the Pacific were known to Europeans more than two centuries before Bilbao crossed the Isthmus of Panama in 1513. Yet though the Spaniards and Portuguese had established important trading posts in Indonesia and the Philippines not a single European vessel sailed into the open Pacific from this side. Winds and currents repelled every attempt. The Pacific Islands only began to be discovered when Magellan sailed west from South America in 1521. Guam in the Marianas, one of the closest islands to Asia, was the first. The European sailing vessels that crossed the Pacific from America were unable to sail back the way they had come and all subsequent discoveries were made on westward voyages. It was not until 1565 that the only feasible return route was discovered, north with the Kuroshio Current off Japan and east in the high latitudes between Hawaii and the Aleutian Islands with the West Wind Drift. The so-called 'caravel route', which resulted from these discoveries and which was followed for more than two centuries, was thus a vast downwind, downstream circuit encompassing the whole of the subtropical gyre.

The areas and seasons of locust breeding are areas and seasons of rain, the summer rainfall regime south of the Sahara contrasting with the winter and spring rains around the Mediterranean and the Persian Gulf. Successive generations of locusts breed in different areas and the seasonal migrations, which take the swarms out of the areas that are beginning to dry up, are accomplished by drift. Individual members of the swarms are usually uniformly orientated in groups but their orientation and locomotion have no significant effect on their displacement by the wind. Downwind movement inevitably carries them into the

Inter-Tropical Convergence Zone and the locusts thus use the kinetic energy of the atmospheric circulation to locate and exploit the temporary but very extensive vegetation which follows the rains in these arid regions.

Whichever of these two parallels is the closer, detailed observations of the tracks of migratory fish on the high seas, accompanied by suitable current measurements, must be awaited with considerable interest.

References

Arnold, G. P. (1974). Rheotropism in fishes. *Biological Reviews* **49**, 515–76.

Berry, J. (1932). Report of an investigation of the migration of smolts in the River Tay during spring 1931. *Salmon Fisheries*, (**VI**), 21 pp.

Calderwood, W. L. (1905). Observations on the seaward migration of smolts in the estuary of the River Tay. *Annual Report of the Fishery Board for Scotland*, (**23**), Appendix III, 80–5.

Creutzberg, F. (1961). On the orientation of migrating elvers (*Anguilla vulgaris* Turt.) in a tidal area. *Netherlands Journal of Sea Research*, **1**, 257–338.

Creutzberg, F., Eltink, A. Th. G. W. & van Noort, G. J. (1978). The migration of plaice larvae *Pleuronectes platessa* into the western Wadden Sea. In *Physiology and Behaviour of Marine Organisms*, ed. D. S. McLusky & A. J. Berry, pp. 243–51. Oxford: Pergamon Press.

Cushing, D. H. (1972). The production cycle and the numbers of marine fish. In *Conservation and Productivity of Natural Waters*, ed. R. W. Edwards & D. J. Garrod, pp. 213–32. Symposia of the Zoological Society of London (**29**). London: Academic Press.

Deelder, C. L. (1958). On the behaviour of elvers (*Anguilla vulgaris* Turt.) migrating from the sea into fresh water. *Journal du Conseil*, **24**, 135–46.

Dodson, J. J. & Leggett, W. C. (1973). Behavior of adult American shad (*Alosa sapidissima*) homing to the Connecticut River from Long Island Sound. *Journal of the Fisheries Research Board of Canada*, **30**, 1847–60.

Dodson, J. J. & Leggett, W. C. (1974). Role of olfaction and vision in the behavior of American shad (*Alosa sapidissima*) homing to the Connecticut River from Long Island Sound. *Journal of the Fisheries Research Board of Canada*, **31**, 1607–19.

French, R., Bilton, H., Osako, M. & Hartt, A. (1976). Distribution and origin of sockeye salmon (*Oncorhynchus nerka*) in offshore waters of the North Pacific ocean. *International North Pacific Fisheries Commission Bulletin*, (**34**), 113 pp.

Fried, S. M., McCleave, J. D. & LaBar, G. W. (1978). Seaward migration of hatchery-reared Atlantic salmon, *Salmo salar*, smolts in the Penobscot River estuary, Maine: riverine movements. *Journal of the Fisheries Research Board of Canada*, **35**, 76–87.

Gibson, R. N. (1973). Tidal and circadian activity rhythms in juvenile plaice, *Pleuronectes platessa*. *Marine Biology*, **22**, 379–86.

Gibson, R. N. (1976). Comparative studies on the rhythms of juvenile flatfish. In *Biological Rhythms in the Marine Environment*, ed. P. J. DeCoursey, pp. 199–213. Columbia: University of South Carolina Press.

Gibson, R. N., Blaxter, J. H. S. & de Groot, S. J. (1978). Developmental changes in the activity rhythms of the plaice (*Pleuronectes platessa* L.). In *Rhythmic Activity of Fishes*, ed. J. E. Thorpe, pp. 169–86. London: Academic Press.

Graham, J. J. (1972). Retention of larval herring within the Sheepscot estuary of Maine. *Fishery Bulletin*, **70**, 299–305.

Graham, J. J. & Davis, C. W. (1971). Estimates of mortality and year-class strength of larval herring in western Maine, 1964–67. *Rapports et Procès-Verbaux des Réunions, Conseil International pour l'Exploration de la Mer*, **160**, 147–52.

Greer Walker, M., Harden Jones, F. R. & Arnold, G. P. (1978). The movements of plaice (*Pleuronectes platessa* L.) tracked in the open sea. *Journal du Conseil*, **38**, 58–86.

Greer Walker, M., Riley, J. D. & Emerson, L. S. (1980). On the movements of sole (*Solea solea*) and dogfish (*Scyliorhinus canicula*) tracked off the East Anglian coast. *Netherlands Journal of Sea Research*, **14**, 66–77.

Groot, C., Simpson, K., Todd, I., Murray, P. D. & Buxton, G. A. (1975). Movements of sockeye salmon (*Oncorhynchus nerka*) in the Skeena River estuary as revealed by ultrasonic tracking. *Journal of the Fisheries Research Board of Canada*, **32**, 233–42.

Groot, S. J. de (1971). On the interrelationships between morphology of the alimentary tract, food and feeding behaviour in flatfishes (Pisces: Pleuronectiformes). *Netherlands Journal of Sea Research*, **5**, 121–96.

Hain, J. H. W. (1975). The behaviour of migratory eels, *Anguilla rostrata*, in response to current, salinity and lunar period. *Helgoländer wissenschaftliche Meeresuntersuchungen*, **27**, 211–33.

Harden Jones, F. R. (1968). *Fish Migration*. London: Arnold.

Harden Jones, F. R. (1979). The migration of plaice (*Pleuronectes platessa*) in relation to the environment. In *Fish behavior and its use in the capture and culture of fishes*, ed. J. E. Bardach, J. J. Magnuson, R. C. May and J. M. Reinhart, pp. 383–99. Manila: International Center for Living Aquatic Resources Management.

Harden Jones, F. R. (1980). The Nekton: production and migration patterns. In *Fundamentals of Aquatic Ecosystems*, ed. R. S. K. Barnes & K. H. Mann, pp. 119–42. Oxford: Blackwell.

Harden Jones, F. R., Arnold, G. P., Greer Walker, M. & Scholes, P. (1979). Selective tidal stream transport and the migration of plaice (*Pleuronectes platessa* L.) in the southern North Sea. *Journal du Conseil*, **38**, 331–7.

Harden Jones, F. R., Greer Walker, M. & Arnold, G. P. (1978). Tactics of fish movement in relation to migration strategy and water circulation. In *Advances in Oceanography*, ed. H. Charnock & Sir George Deacon, pp. 185–207. New York: Plenum Publishing Corporation.

Harden Jones, F. R., Margetts, A. R., Greer Walker, M. & Arnold, G. P. (1977). The efficiency of the Granton otter trawl determined by sector-scanning sonar and acoustic transponding tags. *Rapports et Procès-verbaux des Réunions. Conseil International pour l'Exploration de la Mer*, **170**, 45–51.

Harding, D., Nichols, J. H. & Tungate, D. S. (1978). The spawning of plaice (*Pleuronectes platessa* L.) in the southern North Sea and English Channel. *Rapports et Procès-verbaux des Réunions. Conseil International pour l'Exploration de la Mer*, **172**, 102–13.

Hasler, A. D. (1960). Homing orientation in migrating fishes. *Ergebnisse der Biologie*, **23**, 94–115.

Heyerdahl, T. (1978). *Early Man and the Ocean*. London: Allen & Unwin.

Hughes, D. A. (1969). On the mechanisms underlying tide-associated movements of *Penaeus duorarum* Burkenroad. *FAO Fisheries Report* (57), **3**, 867–74.

Ichihara, T. (1979). Present status of marine biotelemetry in Japan. *Underwater Telemetry Newsletter*, **9**, 5–7.

Ichihara, T., Yonemori, T. & Asai, H. (1975). Swimming behavior of a chum salmon, *Oncorhynchus keta*, on the southern migration off Etorofu Island, the southern Kurile Islands. *Bulletin of the Far Seas Fisheries Research Laboratory, Japan*, (**13**), 63–77. Environment Canada, Fisheries & Marine Service (1977). Translation (3942), 40 pp.

Kalmijn, A. J. (1971). The electric sense of sharks and rays. *Journal of Experimental Biology*, **55**, 371–83.

Kuipers, B. (1973). On the tidal migration of young plaice (*Pleuronectes platessa*) in the Wadden Sea. *Netherlands Journal of Sea Research*, **6**, 376–88.

LaBar, G. W., McCleave, J. D. & Fried, S. M. (1978). Seaward migration of hatchery-reared Atlantic salmon (*Salmo salar*) smolts in the Penobscot River estuary, Maine: open-water movements. *Journal du Conseil*, **38**, 257–69.

Laurs, R. M., Yuen, H. S. H. & Johnson, J. H. (1977). Small-scale movements of albacore, *Thunnus alalunga*, in relation to ocean features as indicated by ultrasonic tracking and oceanographic sampling. *Fishery Bulletin*, **75**, 347–55.

Lawson, K. D. & Carey, F. G. (1972). An acoustic telemetry system for transmitting body and water temperature from free swimming fish. *Woods Hole Oceanographic Institution Technical Report* WHOI-71-67, 21 pp.

Leggett, W. C. (1976). The American shad (*Alosa sapidissima*), with special reference to its migration and population dynamics in the Connecticut River. In *The Connecticut River Ecological Study*, ed. D. Merriman & L. M. Thorpe, pp. 169–225. Washington: American Fisheries Society.

Leggett, W. C. & Trump, C. L. (1978). Energetics of migration in American shad. In *Animal Migration, Navigation, and Homing*, ed. K. Schmidt-Koenig & W. T. Keeton, pp. 370–7. Berlin, Heidelberg: Springer–Verlag.

Mason, J. C. (1974). Behavioral ecology of chum salmon fry (*Oncorhynchus keta*) in a small estuary. *Journal of the Fisheries Research Board of Canada*, **31**, 83–92.

Mather, F. J. (1962). Transatlantic migration of two large bluefin tuna. *Journal du Conseil*, **27**, 325–7.

Meek, A. (1915). Migrations in the sea. *Nature, London*, **95**, 231.

Mitson, R. B. & Cook, J. C. (1971). Shipboard installation and trials of an electronic sector-scanning sonar. *The Radio & Electronic Engineer*, **41**, 339–50.

Mitson, R. B. & Storeton-West, T. J. (1971). A transponding acoustic fish tag. *The Radio & Electronic Engineer*, **41**, 483–9.

Neave, F. (1964). Ocean migrations of Pacific salmon. *Journal of the Fisheries Research Board of Canada*, **21**, 1227–44.

Rainey, R. C. (1978). The evolution and ecology of flight: the 'oceanographic' approach. In *Evolution of Insect Migration and Diapause*, ed. H. Dingle, pp. 33–48. New York: Springer-Verlag.

Rommel, S. A. & McCleave, J. D. (1973a). Sensitivity of American eels (*Anguilla rostrata*) and Atlantic salmon (*Salmo salar*) to weak electric and magnetic fields. *Journal of the Fisheries Research Board of Canada*, **30**, 657–63.

Rommel, S. A. & McCleave, J. D. (1973b). Prediction of oceanic electric fields in relation to fish migration. *Journal du Conseil*, **35**, 27–31.

Royce, W. F., Smith, L. S. & Hartt, A. C. (1968). Models of oceanic migrations of Pacific salmon and comments on guidance mechanisms. *Fishery Bulletin*, **66**, 441–62.

Russell, E. S. (1937). Fish migrations, *Biological Reviews*, **12**, 320–37.

Seckel, G. R. (1972). Hawaiian-caught skipjack tuna and their physical environment. *Fishery Bulletin*, **70**, 763–87.

Sharp, G. D. & Dotson, R. C. (1977). Energy for migration in albacore, *Thunnus alalunga*. *Fishery Bulletin*, **75**, 447–50.

Stasko, A. B. (1975). Progress of migrating Atlantic salmon (*Salmo salar*) along an estuary, observed by ultrasonic tracking. *Journal of Fish Biology*, **7**, 329–38.

Stasko, A. B., Horrall, R. M. & Hasler, A. D. (1976). Coastal movements of adult Fraser River sockeye salmon (*Oncorhynchus nerka*) observed by ultrasonic tracking. *Transactions of the American Fisheries Society*, **105**, 64–71.

Stasko, A. B., Horrall, R. M., Hasler, A. D. & Stasko, D. (1973). Coastal movements of mature Fraser River pink salmon (*Oncorhynchus gorbuscha*) as revealed by ultrasonic tracking. *Journal of the Fisheries Research Board of Canada*, **30**, 1309–16.

Tesch, F-W. (1965). Verhalten der Glasaale (*Anguilla anguilla*) bei ihrer Wanderung in den Ästuarien deutscher Nordseeflüsse. *Helgoländer wissenschaftliche Meeresuntersuchungen*, **12**, 404–19.

Tesch, F-W. (1974). Speed and direction of silver and yellow eels, *Anguilla anguilla*, released and tracked in the open North Sea. *Berichte der Deutschen wissenschaftlichen Kommission für Meeresforschung*, **23**, 181–97.

Thunberg, B. E. (1971). Olfaction in parent stream selection by the alewife (*Alosa pseudoharengus*). *Animal Behaviour*, **19**, 217–25.

Tytler, P., Thorpe, J. E. & Shearer, W. M. (1978). Ultrasonic tracking of the movements of Atlantic salmon smolts (*Salmo salar* L) in the estuaries of two Scottish rivers. *Journal of Fish Biology*, **12**, 575–86.

Veen, J. F. de (1967). On the phenomenon of soles (*Solea solea* L.) swimming at the surface. *Journal du Conseil*, **31**, 207–36.

Veen, J. F. de (1978). On selective tidal transport in the migration of North Sea plaice (*Pleuronectes platessa*) and other flatfish species. *Netherlands Journal of Sea Research*, **12**, 115–47.

Venema, S. C. & Creutzberg, F. (1973). Seasonal migration of the swimming crab *Macropipus holsatus* in an estuarine area controlled by tidal streams. *Netherlands Journal of Sea Research*, **7**, 94–102.

Verheijen, F. J. & Groot, S. J. de (1967). Diurnal activity pattern of plaice and flounder (*Pleuronectidae*) in aquaria. *Netherlands Journal of Sea Research*, **3**, 383–90.

Verwey, J. (1958). Orientation in migrating marine animals and a comparison with that of other migrants. *Archives Néerlandaises de Zoologie*, **13**, Supplement, 418–45.

Voglis, G. M. & Cook, J. C. (1966). Underwater applications of an advanced acoustic scanning equipment. *Ultrasonics*, **4**, 1–9.

Weihs, D. (1978). Tidal stream transport as an efficient method for migration. *Journal du Conseil*, **38**, 92–9.

Westerberg, H. (1979). Counter-current orientation in the migration of the European eel. *Rapports et Procès-verbaux des Réunions. Conseil International pour l'Exploration de la Mer*, **174**, 134–43.

White, H. C. & Huntsman, A. G. (1938). Is local behaviour in salmon heritable? *Journal of the Fisheries Research Board of Canada*, **4**, 1–18.

Williams, F. (1972). Consideration of three proposed models of the migration of young skipjack tuna (*Katsuwonus pelamis*) into the eastern Pacific Ocean. *Fishery Bulletin*, **70**, 741–62.

WILLIAM T.KEETON

The orientation and navigation of birds*

Research on the mechanisms of bird orientation and navigation is currently experiencing a period of very rapid development, with the result that most
summaries of this field in the popular literature are badly out of date. My purpose
here is to provide the reader with some understanding of modern-day ideas and
research thrusts in this field, and to try to communicate a sense of the intense
excitement that now prevails in this branch of ornithology.

The more familiar orientational cues

One of the earliest and most stimulating discoveries in the field of avian
orientation and navigation was that celestial cues – the sun and the stars – play
a fundamental role in bird orientation. Thus, more than twenty-five years ago,
Gustav Kramer (1952) and his students showed that birds possess what has come
to be called a sun compass. And, following up Kramer's (1949, 1951) early
observations that several species of nocturnal migrants exhibit oriented migratory
restlessness (*Zugunruhe*) in circular cages under clear night skies, Sauer (1957)
found that such nocturnal orientation depended on the stars, thus laying the basis
for what is now known as the star compass.

The overwhelming preponderance of evidence now indicates that solar cues
are used by birds only as a simple compass, not as the basis of a bicoordinate
navigation system as proposed by Matthews (1953, 1955) in his imaginative and
historically important sun-arc hypothesis (for a summary of the evidence, see
Keeton, 1974a). In other words, it appears to be only the sun's azimuth (direction from the observer) that provides the bird with orientational information. But
the azimuth can provide such information only if the time is known, so that
compensation can be made for the sun's changing position throughout the day.
That birds do indeed couple their internal clock (circadian rhythm) with their
observation of the sun's azimuth in determining compass directions was clearly

*This article was first published in *British Birds* (1979, volume **72**, pages 451–70), and is here reprinted (with minor up-dating) with per-mission. It is a shortened version of a paper to be published in the Proceedings of the XVIIth International Ornithological Congress, Berlin. The author's research was supported by the National Science Foundation.

demonstrated by Kramer (1953a) and by Hoffmann (1954), working with caged birds. Later, Schmidt–Koenig (1958, 1960) extended this finding to free-flying homing pigeons. He showed that pigeons whose internal clocks had been experimentally shifted six hours out of phase with true sun time chose initial bearings roughly 90° from those of control pigeons when released at a distant test site (Fig. 1a); their clocks had been shifted a quarter of a day, and, as a consequence, they misread the sun compass and chose bearings a quarter of a circle different from those of the controls.

As in the case of the sun, the evidence is overwhelming that the stars provide only compass information for birds, even though they could potentially be used in true bicoordinate navigation (for a thorough discussion, see Emlen, 1975a). Sauer (1957) thought that the star compass, like the sun compass, required time

Fig. 1. Vanishing bearings of pigeons that have been clock-shifted 6 hours fast. (a) Experienced pigeons released on a sunny day at a distant site. The mean bearing of the clock-shifted birds is roughly 90° to the left of that of the controls. (b) Experienced pigeons released on a sunny day at a site less than a mile from home, where the landscape should be completely familiar. Again, the mean bearing of the clock-shifted birds is roughly 90° to the left of that of the controls. (c) Experienced pigeons released on a totally overcast day at a distant unfamiliar site. Both the clock-shifted and the control birds are homeward oriented, and there is no indication of a difference between them, which suggests that, in the absence of the sun compass, the pigeons use orientational cues that do not require time compensation. In this and later figures showing bearings, north is indicated by a small line at the top of the circle, and the home direction by a dashed line reaching the perimeter of the circle. The bearing of each individual bird is shown as a small symbol on the outside of the circle; where two treatments are included on a single circle, the bearings of the controls are shown as open symbols and the bearings of the experimental birds as filled symbols. The mean vectors are shown as arrows (with open or filled heads, respectively), whose length is drawn proportional to the tightness of clumping of the bearings (i.e. the longer the vector – at maximum reaching the perimeter of the circle – the better oriented the sample of bearings). The uniform probability under the Rayleigh test is given inside the circle; the first value is for the controls and the second for the experimentals.

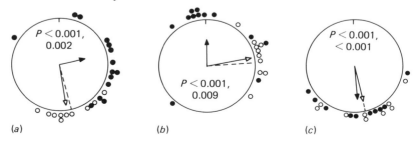

(a)   (b)   (c)

compensation. But in a detailed study of the migratory orientation of the Indigo Bunting (*Passerina cyanea*), Emlen (1967) found that this species uses star patterns to determine directions, a process that does not require time compensation (Fig. 2). It seems likely that this is true of other species as well (Emlen, 1975*a*). Thus the way birds read the star compass differs fundamentally from the way they read the sun compass.

Although the sun compass during the day and the star compass during the night are certainly dominant orientational cues for birds, there is now abundant evidence that neither is essential for proper orientation. Thus experienced homing pigeons can orient accurately homeward from distant unfamiliar release sites under heavy total overcast (Fig. 1*c*) (Keeton, 1969, 1974*a*). And tracking-radar studies have regularly revealed nocturnal migrants in oriented flight under heavy overcast when the stars are not visible (e.g. Nisbet & Drury, 1967; Steidinger, 1968; Williams, Williams, Teal & Kanwisher, 1972; Griffin, 1972, 1973). It is apparent, then, that though birds often use celestial cues when available, they can use alternative cues when necessary. In short, avian orientation systems include redundant, or back-up, cues (Keeton, 1974*a*; Emlen, 1975*a*).

It is also important to point out, as Kramer (1953*b*) did long ago, that a compass alone cannot tell a bird where it is or which direction it should fly to reach

Fig. 2. An example of how north can be located by star patterns. If one draws an arrow running through two particular stars in the cup of the big dipper (Ursa Major), it will point toward Polaris. Although the position of the constellations changes during the night (two positions are shown here), the same stars always determine an arrow pointing toward Polaris (i.e. toward north, or the pole of the celestial rotation), hence directions can be determined without need of time compensation. Many different star patterns could be used for direction finding in this way.

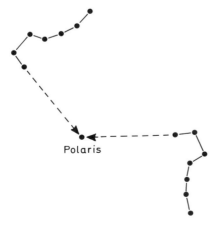

Polaris

a particular destination. Other environmental cues must provide goal-orienting birds with an analog of map information. What might these cues be? Certainly the most obvious possibilities would be familiar landmarks, but, curiously enough, birds seem to make only minimal use of these. Radar studies strongly suggest that landmarks play little role in migratory orientation (Emlen, 1975a; Emlen & Demong, 1978). Moreover, pigeons clock-shifted six hours out of phase with true sun time and released half a mile from their home loft, in an area they have flown over every day when exercising, usually choose bearings deflected 90° from the homeward course (Fig. 1(b)) (Graue, 1963; Keeton, 1974a); the birds act as though they have never before seen the place! Even more remarkable, pigeons wearing frosted-white contact lenses that eliminate image vision beyond two or three meters can orient accurately homeward from distant locations, and can even tell when they have arrived at their goal (Schmidt–Koenig & Schlichte, 1972; Schlichte, 1973).

Another possibility that comes to mind in this age of rocketry is that the birds might have some sort of sophisticated inertial guidance system that would permit them to detect all the angular accelerations of the outward journey and then double integrate them to determine the return course (Barlow, 1964). Although this possibility cannot be completely ruled out, it seems very unlikely in view of evidence against it from many kinds of investigations (summarized in Keeton, 1974a).

Unusual sensory capabilities of birds

Having eliminated the most obvious cues, we turn next to special sensory capabilities (some of them only very recently discovered) that may play a role in the birds' amazing navigational feats.

Magnetic detection

Although it had often been suggested that birds might be able to derive directional information from the earth's magnetic field, the prevailing opinion of this possibility in the scientific community of the mid 1960s was one of intense skepticism. Most investigators thought it very unlikely that any organism could detect a magnetic field as weak as that of the earth (about 0.5 Gauss). But beginning in the mid 1960s, and continuing to the present, a group in Frankfurt, led first by F. Merkel and later by W. Wiltschko, intensively investigated the possible role of magnetic cues in avian orientation.

Having first found that European robins (*Erithacus rubecula*) exhibit migratorily appropriate orientation in circular test cages when visual cues are unavailable (Merkel, Fromme & Wiltschko, 1964), this group went on to show that the orientation of the robins could be changed in a predictable way by turning the magnetic field (e.g. making magnetic north in the cage coincide with geographic east), using Helmholtz coils positioned around the test cage (Merkel & Wiltschko, 1965; Wiltschko, 1968). Their results were later successfully replicated by Wallraff (1972) with robins, and by Emlen *et al.* (1976) with indigo buntings.

Wiltschko (1972; see also Wiltschko & Wiltschko, 1972) found, further, that robins apparently pay no heed to the polarity of the magnetic field, but rather, in the northern hemisphere, take north as that direction in which the magnetic and gravity vectors form the most acute angle (Fig. 3). They are unable to orient in visually cueless test cages when the magnetic field is entirely horizontal (i.e. has no vertical component), as is the case at the equator. In short, the birds' manner of reading the magnetic compass is very different from our own.

The first good evidence of a magnetic effect on pigeon homing came when I (Keeton, 1971, 1972) found that bar magnets attached to experienced birds' backs often cause disorientation on heavily overcast days, whereas they had little effect on sunny days. Walcott & Green (1974), using Helmholtz coils on the pigeons' head and neck, exposed the birds to more homogeneous magnetic fields than those produced by my bar magnets; again there was no dramatic effect on

Fig. 3. The magnetic compass of the European robin. Top: The birds orient northward in spring, whether the magnetic field vector points north and down (which is the normal condition) or south and up. Bottom: The same birds change their orientation to southward if the magnetic vector points north and up or south and down. In short, it appears to be the alignment of the magnetic vector, not its polarity, that determines the birds' behavior.

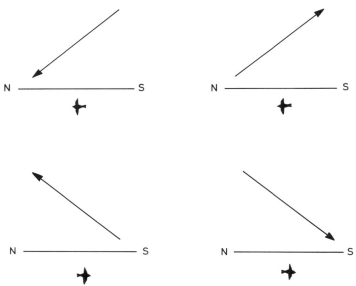

sunny days, but on overcast days the birds' orientation was changed (not merely disrupted) in a manner consistent with Wiltschko's formulation of the way the avian magnetic compass works. These results suggest strongly that the disorienting effects of bar magnets I had found earlier were not due merely to some general physiological disturbance.

Thus emerged the concept that, in the case of homing pigeons, the magnetic field provides compass information that experienced birds use primarily when the sun compass is not available. While probably generally true, this formulation may be an oversimplification. There is a growing body of evidence that magnetic perturbations alter in a small but consistent way the orientation of pigeons on sunny days (Keeton, 1971; Keeton, Larkin & Windsor, 1974; Larkin & Keeton, 1976; Walcott, 1977). Moreover, there is also some indication that magnetic information obtained during the outward journey to the release site may sometimes influence the initial orientation of pigeons (R. Wiltschko, Wiltschko & Keeton, 1978; Kiepenheuer, 1978; Papi et al., 1978a). It therefore seems possible that magnetic cues may play some role in the map aspect of avian navigation.

Lindauer & Martin (1968; see also Martin & Lindauer, 1977) have found that honeybees are sensitive to magnetic changes of less than 10^{-3} Gauss and very probably of less than 10^{-5} Gauss. There is now evidence from several sources that birds are probably equally sensitive. Thus Southern (1972; but see also 1978) has reported for gull chicks, Keeton et al. (1974; see also, Larkin & Keeton, 1977) for homing pigeons, and F. Moore (1977) for free-flying migrants that orientation is influenced by natural magnetic disturbances, due largely to events on the sun, such as solar flares. As B. Moore (1980) has pointed out, the data of Keeton et al., (1974) suggest that magnetic fluctuations of only one milligauss – roughly a five-hundredth of the earth's natural field – can result in rotations of 10°–40° in pigeons' choice of bearings. Also indicating a very great magnetic sensitivity is evidence reported by Wagner (1976) and by Walcott (1978) that geographic magnetic anomalies may disturb the initial orientation of pigeons.

In view of the abundant evidence that birds are very sensitive to magnetic stimuli, at least when they are orienting, one would hope that we will soon learn the mechanism of their magnetic sense. The recent discovery by Walcott, Gould & Kirschvink (1979) of the ferromagnetic mineral magnetite in the heads of pigeons, followed by the finding of Presti & Pettigrew (1980) of what is probably the same material in the necks of pigeons and White-crowned Sparrows (Zonotrichia leucophrys), is very exciting, but it remains to be determined whether this material functions in magnetic detection. Evaluation of this and other possibilities should be facilitated by Bookman's (1977) reported success in training pigeons in a two-choice test where different magnetic conditions provide the information on which the choice must be based.

Possible detection of gravity variations

Several years ago, T. Larkin and I noticed that often when animals have shown clear responsiveness to magnetic stimuli they have been simultaneously responding to gravity (e.g. Lindauer & Martin, 1968; Wehner & Labhart, 1970; Wiltschko & Wiltschko, 1972). Consequently, we sought to determine whether gravity cues might play some role in pigeon homing, especially in situations where the birds are using magnetic information. Being unable to alter gravity in the laboratory, we looked instead for a possible influence of the natural monthly gravitational cycle caused by the changing relative positions of the earth, sun, and moon. We soon found a significant correlation between the pigeon's mean vanishing bearings and the day of the lunar synodic month (Larkin & Keeton, 1978). Suggestive as these results may be, however, they do not prove a direct effect of gravitational changes on the birds' orientation, because some other environmental variables to which birds might be responsive may also be related to the lunar cycle.

Despite the still uncertain meaning of the synodic lunar rhythm we found, let us consider the possibilities for birds if they could detect minute variations in gravity. Gravity varies not only temporally but also geographically, in both a regular and an irregular manner. The regular variation is in a north–south gradient and is due to the fact that the earth is not a perfect sphere. The irregular variation is due to the differing densities of the material in the earth's crust at different localities. Gravity cues could, then, potentially be useful in navigation, both because they could indicate the north-south axis and because they could provide another topography in addition to those we normally consider.

If a bird could use the north-south gravitational gradient to determine true (i.e. rotational) north, then magnetic declination (the deviation of magnetic north from true north) might be readable. It happens that declination is one of the very few environmental parameters that vary as a rough analog of longitude, hence its potential usefulness would be very great indeed. Moreover, some of our preliminary results (Larkin & Keeton, in preparation) suggest that declination may actually be the parameter of the magnetic field that most influences pigeon orientation during magnetic disturbances. It is important to emphasize, however, that there is as yet no direct evidence that birds can detect such incredibly tiny differences in gravity (less than ten Gal) as would be necessary to permit use of gravitational cues in long-distance navigation.

Barometric pressure detection

For birds, which spend much of their time in the air, a sensitive ability to detect changes in barometric pressure would potentially be useful in a variety of ways. Hence it is not surprising that pigeons have recently been shown to possess just such a detection capability (Kreithen & Keeton, 1974a; Delius & Emmerton, 1978). Other species of birds have not yet been tested for this ability.

One obvious way birds could use a barometric pressure sense is as an altimeter. The sensitivity found in homing pigeons is sufficiently great that they should surely be able to detect a change in altitude of ten metres; indeed it seems likely they can detect much smaller changes. Not only would an altimeter sense be useful when flying in cloud, but we note also that, if birds really possess the ability to detect changes in gravity, as speculated above, an altimeter would be essential because compensation for altitude would be necessary in view of the fall in gravity with increasing elevation.

Human meteorologists find a sensitive barometer very helpful in predicting weather changes; why should not a bird use its built-in barometer in the same way? There is, of course, a long history of reports from bird watchers that the birds in their garden seem to show by behavior changes that a weather front is approaching, long before the human observer sees any indication of the front. And there is convincing recent evidence that birds about to initiate migratory flight are very good meteorologists indeed (Emlen, 1975a). Thus, in eastern North America, major autumnal movements tend to occur on the east side of a high-pressure cell following passage of a cold front, and major spring flights tend to concentrate on the west side of a high-pressure area ahead of an advancing low-pressure cell (Richardson, 1978). Many migrants, especially small song birds, seem to be quite accurate at predicting early in the evening, when they are still on the ground, what the wind conditions aloft will be later that night; they go up in greatest numbers when the winds aloft will be favorable.

Finally, the last few years have witnessed a growing interest in the possibility that weather factors such as wind directions, pressure patterns in the atmosphere, and patterns of air turbulence could potentially provide useful orientational information (Griffin, 1969; Emlen, 1975a), and the detection of these would surely be facilitated by the birds' barometric sense.

Infrasound detection

Although the literature would lead one to think the lower frequency limit of bird hearing would be about 100 Hz, Yodlowski, Kreithen & Keeton (1977) have found that homing pigeons are sensitive to frequencies lower than 1 Hz. In fact, the audiogram for the pigeon worked out by Kreithen & Quine (1979) shows sensitivity down at least to 0.05 Hz. The pigeons' sensitivity is at the appropriate level for extracting meaningful signals from the background environmental noise.

Detection of infrasound by birds raises an intriguing new orientational possibility. Because attenuation is proportional to the square of frequency, infrasonic frequencies as low as those the birds can detect may travel hundreds or even thousands of miles with little energy loss. Potentially, then, pigeons might be able to monitor distant infrasonic sources, such as mountains whistling because of winds blowing across them, and to use these as a rough system of beacons for determining position.

But if pigeons are to use infrasounds in this way, they should be able to tell from what directions the sounds come, yet binaural comparisons would be impossible for such very long wavelengths (over 3 km at 0.1 Hz). A possible way around this problem comes from the work of Quine (1979) who finds that pigeons' frequency discrimination in the infrasonic range is sufficiently good so that doppler shifts in apparent frequency induced when a bird flies toward and then away from an infrasound source would be well within the birds' detection range. Thus pigeons may be able to get directional information from infrasounds while in flight, even if they cannot do so while perching. Field experiments to evaluate this possibility are now in progress at Cornell.

Polarized light detection

The discovery by Kreithen & Keeton (1974*b*) and by Delius, Perchard & Emmerton (1976) that homing pigeons can detect the polarization of light may mean that these birds, like honeybees, can continue using a derivative of the sun compass on partially overcast days, when the sun's disk is hidden from view but some blue sky remains. This is possible because of the geometric relationships between the position of the observer, the plane of polarization of sunlight, and the position of the sun, which permits derivation of the sun's position if the polarization can be detected. In addition, there are other more direct ways in which sky-light polarization may be used in orientation; thus, for example, the band of maximum polarization rotates around the celestial north pole during the course of each day, so if the rotation could be observed then true north could be determined directly (Brines, in press). Actual use of polarized light by orienting birds in the field has, however, not yet been studied.

Ultraviolet light detection

Yet another addition to our knowledge of avian vision is the recent discovery by Kreithen & Eisner (1978) that homing pigeons can see ultraviolet light (see also Delius & Emmerton, 1978). This raises the question, now under investigation at Cornell, whether the pigeons ordinarily perform their analysis of polar polarization in the ultraviolet wavelengths, as honeybees do.

Olfaction

Olfaction, unlike most of the other sensory capacities discussed above, is not a recently discovered sense. But the possibility that it may play an important role in avian navigation is new. It was in 1972 that Papi, Fiore, Fiaschi & Benvenuti first put forward their olfactory navigation hypothesis. Briefly, they propose that young pigeons at the home loft would learn to associate particular odors with winds from certain directions. Thus odor A might arrive at the loft primarily on winds from the north, odor B on winds from the east, etc. A bird released at a distant site, say north of home, would detect a strong odor of A and therefore determine its position to be north of home. The bird would then use one of its compass systems to locate south, and begin its homeward flight.

Papi and his colleagues have performed a long series of ingenious experiments to test their hypothesis, and have reported consistently positive results (for summary, see Papi, 1976). Unfortunately, attempts to repeat some of these experiments at Cornell yielded generally negative results as far as orientation was concerned, though there was sometimes an effect on homing success (Keeton, 1947b; Keeton & Brown, 1976; Keeton, Kreiten & Hermayer, 1977; Hermayer & Keeton, 1979). Similar attempts at Tübingen in Germany yielded either negative results (Schmidt–Koenig & Phillips, 1978) or ambiguous ones (Hartwick, Kiepenheuer & Schmidt–Koenig, 1978).

In an effort to resolve the differences between the Pisa and Cornell groups, the two groups performed six series of collaborative experiments at Cornell in 1977. The overall result of these experiments was that, with the exception of deflector-loft experiments, to be discussed later, we found no consistent effect of olfactory interference or deprivation on initial orientation, but we did often find an effect on homing success from distant unfamiliar release sites (Papi et al., 1978a; Papi, Keeton, Brown & Benvenuti, 1978b). Unfortunately these results are interpreted one way by the Cornell investigators and another way by the Italian investigators; hence the question of the role of olfaction in avian orientation remains unresolved. Papi and his colleagues feel the poor homing means that olfactory cues play an irreplaceable role in homing from unfamiliar sites. My colleagues and I feel, on the other hand, that since the olfactorily deprived birds usually depart from the release site on a proper course, their poorer homing may have nothing to do with navigation but may merely indicate a diminished motivation that results in the birds' landing when only part way home. We are concerned about the motivational effects of procedures that interfere in any way with the respiratory system, on which flying birds must make very heavy demands. We attempted to evaluate our suggestion by airplane-tracking pigeons in one of the collaborative experiments, but unfortunately, our suggestion proved almost too true – all five of the experimental birds that were tracked soon landed and we were unable to get as much detail about the flight course as we had hoped. However, before they landed, these birds were flying roughly the same course followed by control birds.

In one series of collaborative experiments we were successful in repeating the Italian group's results on initial orientation. This series utilized the deflector-loft technique so imaginatively designed by Baldaccini, Benvenuti, Fiaschi & Papi (1975). In these experiments, pigeons were exposed in their home lofts to winds

– and the odors they are presumed to carry – deflected either clockwise or counterclockwise (Fig. 4). When tested at release sites, the pigeons chose bearings to the right or left of control pigeons, as predicted by the olfactory hypothesis (Waldvogel, Benvenuti, Keeton & Papi, 1978). These results may indicate that olfactory cues are sometimes used by the Cornell pigeons, but no final decision can be made until experiments that control for the various other orientationally relevant factors altered by the deflectors are completed.

In summary, my appraisal of the role of olfaction in avian navigation is that it is likely that odors constitute one of the many sources of information birds may use in navigating. Olfactory cues appear to be used more by Italian than by Cornell pigeons, for reasons yet to be determined. This is not the first example of major geographic difference in bird orientation. Nocturnal passerine migrants in the southeastern United States nearly always fly downwind (Gauthreaux & Able, 1970; Able 1973, 1974a) whereas the same species in the northeastern

Fig. 4. The deflector loft experiments. Top: All three lofts (control loft in center, lofts with deflectors on each side) have walls that allow free flow of air. Winds from the north, presumably carrying odor A, enter the control loft from the North, but they enter the other lofts from the east and west because of the deflectors. Bottom: Bearings of pigeons released north of home. The control birds, which had earlier experienced normal air flow (i.e. A winds from the north) oriented properly southward, toward home. By contrast, the CW birds, which had experienced A winds from the east, oriented more westerly, and the CCW birds, which had experienced A winds from the west, oriented easterly. (Redrawn from Baldaccini et al., 1975b.)

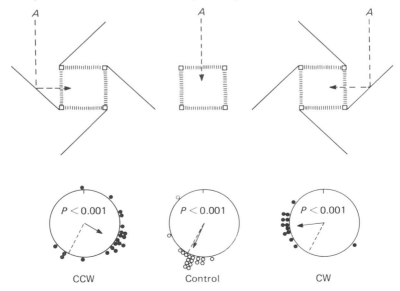

part of the country do so only if the wind direction is migratorily appropriate (Able, 1974*b*, 1978). However, I doubt that olfaction will be found to play as essential a role in avian navigation as Papi and his colleagues have proposed. Indeed, I am not convinced that any single cue so far discovered is essential; there seems to be so much redundancy in the avian navigation system that experienced birds can orient when only a few of the many possible cues are available.

The integration of orientational cues

From all that has been said above, it should be clear to the reader that birds can integrate the many orientational cues they use in a variety of ways – according to different weighting schemes, if you will – depending on the birds' age, experience, and species, and on weather conditions, the season of the year, and the geographic location. The old hope that a single system would be found to explain all of avian orientation has gone aglimmering. Hence one of the chief thrusts of current orientational research is the attempt to learn how the different cues are integrated and what constraints there are on the amazing flexibility of avian navigation systems.

Studies of integration: the ontogeny of orientational behavior

One powerful way of teasing apart the many elements of avian orientation systems is to manipulate the early development of those systems. By so doing, one can often get young birds to omit one or more of the usual cues or to adopt an atypical weighting scheme so that cues that would normally be secondary or tertiary become primary and thus easier to study.

In a series of experiments on the ontogeny of stellar orientation in indigo buntings, Emlen (1972) found that there is a sensitive phase during which young buntings learn to read the star compass. If the buntings have had a view of the starry sky during the weeks preceding the start of their first autumnal migration season, they can orient properly when that season begins. But if they have not seen the night sky until after the first migration season has begun, they never learn to use the star compass, no matter how often they see the sky thereafter.

Emlen (1972, 1975*b*) also showed that young buntings respond initially to the apparent rotation of the starry sky during the night. The axis of this rotation is north-south, hence it can provide compass information. But the birds do not long depend on the axis of rotation *per se;* rather they soon learn star patterns that will indicate where the axis is, and thereafter they rely exclusively on those patterns. In other words, the axis of rotation functions in ontogeny only as the reference against which the star compass is initially calibrated. When Emlen exposed hand-reared young buntings to a planetarium sky rotating around an incorrect axis, the buntings learned to use star patterns appropriate to that axis, and consequently they oriented in an inappropriate direction when later tested under a normal sky. When retested a year later, after extensive exposure to the normal sky, they had not corrected their orientation; what they had learned during the sensitive phase in their early life still dominated their behavior.

In orientation studies with young first-flight pigeons (i.e. very young birds released for their first homing flight), we have found that these birds appear to require the sun compass for orientation; if they are released under heavy total overcast they usually depart randomly (Keeton & Gobert, 1970), even though experienced pigeons can orient accurately under such conditions (Keeton, 1969). Moreover, first-flight youngsters appear to require magnetic cues also; they usually depart randomly, even on sunny days, when wearing bar magnets attached to their backs (Fig. 5) (Keeton, 1971). In short, the first-flight birds need both sun and magnetic cues, whereas experienced pigeons need only one or the other. It seems, then, that these young inexperienced pigeons are integrating cues in a manner quite different from that used by experienced birds. Perhaps the effect of experience is merely to enable them to get by with less information, or alternatively, the experience may help them establish a hierarchy of choice, so that they can later deal with situations in which two or more cues give conflicting information.

As an example of an experimental manipulation that results in young birds' omitting a normal cue, we can cite the studies of Wiltschko, Wiltschko, Brown, & Keeton (in preparation) with so-called no-sun pigeons. In these studies young pigeons raised without ever having a chance to see the sun (they were flown for exercise only on overcast days) were found to be able to orient perfectly well when released under total overcast for their first homing flight, even though normal first-flight pigeons vanish randomly under total overcast (Keeton & Gobert, 1970). Having never viewed the sun, the no-sun pigeons had not incorporated it into their navigation system and hence had no difficulty orienting when it was missing.

Studies of the ontogeny of the sun compass in pigeons have revealed that the coupling of times, directions, and sun azimuths is not inherited but must be

Fig 5. Bearings on a sunny day of first-flight pigeons wearing magnet bars or brass bars. The magnet-laden pigeons (black symbols) vanished randomly, whereas the brass-laden birds (open-symbols) were well oriented.

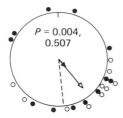

$P = 0.004$, 0.507

learned (Wiltschko, Wiltschko & Keeton, 1976). Thus young pigeons raised under a permanently six-hours-slow clock-shifted photoperiod orient normally, with no indication of the deflection seen in ordinary clock-shifts (Fig. 6a). The birds appear to have learned that the 'morning' sun is in the south, the 'noon' sun is in the west, etc. When these birds are moved to a normal photoperiod and retested after five or six days, they then show bearings deflected 90° from the controls (Fig. 6b); being put in a normal photoperiod has the same effect on them as a six-hours-fast clock-shift has on normal pigeons. These results indicate that the sun compass must be calibrated, which suggests that it may be a derivative compass – that there may be some other more fundamental directional cue that functions as the reference for calibration. One current line of research is the attempt to determine what that reference cue might be.

Studies of integration: manipulation of physiological condition

Emlen (1969) has pursued the question whether it is the seasonal differences in the temporal positions of the stars that determines southward orientation by migratory birds in autumn and northward orientation in spring, or whether the differences in orientation result from corresponding differences in the physiological condition of the birds. By manipulating photoperiods, he contrived to bring one group of male indigo buntings into autumnal condition at the same time as another group was in spring condition. He then tested both groups simultaneously under a spring sky in a planetarium. The birds in autumnal condition oriented southward, whereas the birds in spring condition oriented northward (Fig. 7). Since the two groups saw identical star patterns, Emlen concluded that their different directions of orientation were due to their physiological conditions, not to the environmental stimuli. He predicted that the important factor would be found to be hormonal. Later studies by Martin & Meier (1973) appear to support Emlen's prediction by suggesting that the orientation of white-throated sparrows

Fig. 6. Bearings of pigeons subjected to a 'permanent' six-hour clock-shift. Left: while still living under the shifted photoperiod, the experimental birds orient like the controls, toward home. Right: when retested 5 days after being moved to the normal photoperiod, the experimental birds choose bearings deflected clockwise from those of the controls. (From Wiltschko *et al.*, 1976.)

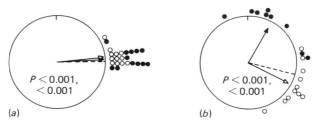

(*Zonotrichia albicollis*) in circular cages can be reversed by altering the temporal pattern of administration of prolactin and corticosterone.

Studies of integration: conflicts between cues

Another method for getting at the question of the relationship between cues is to pit one cue against another in orientational experiments. Let us examine a few recent examples of investigations using this approach. Wiltschko & Wiltschko (1975) have conducted circular-cage experiments in which migratory birds are able to see the starry sky while experiencing a magnetic field that has been turned by Helmholtz coils. In other words, the birds receive conflicting information from the star and magnetic compasses. The Wiltschkos report evidence that their birds (both European robins and warblers) periodically use the magnetic field to recalibrate their star compass. The birds may then orient by the stars for a day or so before again taking a magnetic reading. The process is similar to one a person might use if, after consulting his magnetic compass, he then walked toward a distant tree seen to be in the desired direction; the person might not take another magnetic bearing until he needed to recheck the visual marker or to choose a new one. Wiltschko & Wiltschko (1976) also report that the birds can use the magnetic field to calibrate an entirely artificial 'star' pattern.

It is important to note the difference between these results of the Wiltschkos, in which magnetic cues appear to be used to calibrate the star compass, and the results of Emlen (1970, 1972), in which the axis of celestial rotation is used. It is possible that the main difference is that Emlen was studying the original calibration of the compass by premigratory young birds, whereas the Wiltschkos were testing recalibration by actively migrating birds. Nonetheless, it remains unclear why Emlen's buntings, which have often been tested in a planetarium

Fig. 7. The orientation under a spring planetarium sky of indigo buntings in spring and 'autumnal' physiological conditions. The birds in spring oriented north-northeastward, the usual direction for this species. The birds brought artificially into autumnal condition oriented south-southeastward. (Redrawn from data in Emlen, 1975*a*.)

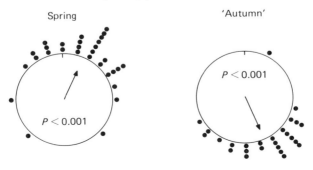

with the celestial axis not aligned along magnetic north-south, have shown no indication of recalibrating their star compass.

Several series of experiments have been conducted at Cornell in which homing pigeons being clock-shifted six hours are permitted exposure to the sun and other natural cues during the shifting process. Adult pigeons allowed to sit in a wire aviary during the overlap period between their shifted day and the real day gave no indication that they derived any orientationally meaningful information from the sun except that the light was on; the timing information potentially available in the sun's position on its arc appeared to be ignored (Alexander & Keeton, 1974).

Extending this approach a step farther, J. R. Alexander & W. T. Keeton (in preparation) tried letting pigeons in the process of being clock-shifted fly for exercise during the overlap period between the shifted and real days. Now there was an effect on orientation. When tested at a distant release site, young birds given this treatment chose initial bearings deflected less than the expected 90° from those of the control birds (the effect differed somewhat depending on the birds' ages). Apparently flight, as opposed to merely sitting in an aviary, had made the birds more responsive to the conflict between their sun compass, their magnetic compass, their visual contact with the loft, and other relevant cues, with the result that the birds given exercise flights during time shifting had altered the way they integrated the various cues. Edrich & Keeton (1978), investigating how much flight is required to produce this effect, found that even pigeons exercised only in flight cages 7.3 m long later chose bearings deflected less than those of normally clock-shifted birds. Further experiments are needed to clarify what cues the clock-shifted birds are using to correct for the erroneous information they get from the sun compass.

Studies of integration: radar tracking of known individuals

In the past, most radar studies of bird migration have used surveillance radar to monitor migratory movements. While this approach has yielded invaluable information concerning flight paths and concerning the relationships between migratory intensity and accuracy and the weather, it has not been suited to study of the orientation of known individual birds. However, Emlen & Demong (1978) have recently used a large tracking radar at the Wallops Island, Virginia, NASA base to follow individual white-throated sparrows previously captured and assayed for stage of molt, amount of fat, and intensity of *Zugunruhe*. The sparrows could be released aloft under various weather conditions (including ones when the birds would not normally begin a migratory flight) and after they had been subjected to various manipulations, such as clock-shifting or having a magnet attached. It was thus possible for Emlen & Demong to investigate the decision-making stage of free-flying migration under conditions when different weightings of orientational cues might be expected.

Amongst their many results, Emlen & Demong (1978, and in press) found that information from viewing the setting sun probably plays an important integrative role during the transition between daylight and darkness. Under overcast skies at night, birds that had not seen sunset oriented poorly whereas those that had seen sunset oriented well. This discovery raises a host of new questions for further investigation. For example, might sunset function as a reference for calibration of other cues? And if so, which cues?

Studies of integration: release-site biases

Many investigators have noticed that each release site used for homing pigeons can be characterized by a preferred departure direction that often deviates somewhat from the true home direction; this release-site bias may be only a few degrees to the left or right of home, or at some locations it may be as much as 60–80 degrees (Keeton, 1973; Wallraff, 1978). Curiously, the bias is usually not much affected by the previous experience of the birds (Fig. 8a, b, c). It remains on overcast days (Fig. 8d) and when the pigeons are wearing magnets (Fig. 8e); hence the bias is probably not a function of either the sun or magnetic compasses. Tests with frosted contact lenses indicate that the bias is not due to the birds' visual perception of the surrounding landscape (Fig. 8f). The preferred direction is not altered by application of α-pinene to the birds noses (Fig. 8g), hence it seems unlikely that the bias has anything to do with olfactory position-determination in the manner proposed by Papi et al. (1972). Pigeons from other lofts often choose bearings at these sites that are comparably deflected from their own home directions (Fig. 8h). Bank swallows (Riparia riparia) show similar biases at these sites, so the biasing factor (or factors), whatever it may be, is apparently not unique to pigeons (Keeton, 1973).

The hope has been that studies of release-site biases (or preferred compass directions, as Wallraff (1978) prefers to call them) would help reveal local factors that might be at least a part of the long-sought navigational map, and would show how those factors are integrated with the orientational cues already known. So far, unfortunately, that hope has not been realized. All efforts to explain the biases have failed. Nonetheless, this approach seems worth continued effort.

Concluding comments

From the above account, it should be apparent to the reader that avian orientation and navigation is one of the most active fields of ornithological

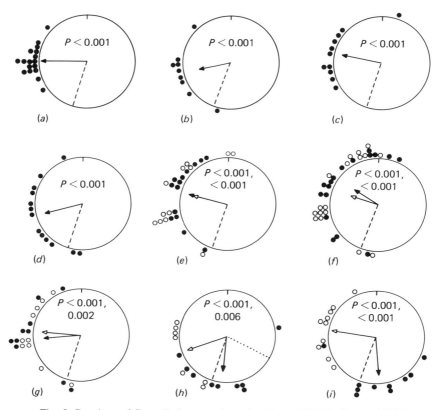

Fig. 8. Bearings of Cornell pigeons released at Castor Hill Fire Tower (143 km NNE of the loft). (a) Experienced birds new-to-site customarily depart in a direction 60 to 80 degrees clockwise (westerly) from home. Home line is indicated by dashes. (b) Birds with prior releases from this site usually continue to choose bearings markedly westerly from the home direction. (c) First-flight youngsters also choose westerly bearings. (d) Even under total overcast, experienced birds new-to-site depart in a westerly direction. (e) The bearings of experienced birds new-to-site are nearly identical, whether the birds are wearing magnet bars (black symbols) or brass bars (open bymbols). (f) Experienced birds new-to-site choose westerly bearings, whether they are wearing frosted-white contact lenses (black symbols) or clear lenses (open symbols). (g) The bearings of experienced birds new-to-site whose beaks and noses had been painted with α-pinene in vasoline (black symbols) did not differ significantly from those of control birds treated with plain vasoline (open symbols). (h) The bearings of pigeons brought from Schenectady, New York (black symbols), were deflected clockwise from *their* home direction (dotted line) in a manner similar to the deflection of the bearings of Cornell birds (open symbols) from the direction to Ithaca (dashed line); it therefore appears that some cue basic to pigeon navigation is rotated clockwise at Castor Hill, and that pigeons with different destinations read the cue in the same manner. (i) Although pigeons clock-shifted six hours fast choose bearings (black symbols) that are more nearly homeward oriented than those of normal pigeons (open symbols), their homing success is considerably poorer.

research today. Our whole way of thinking about the subject has changed radically in less than a decade, and the change continues unslowed. A host of new cues are being discovered, and a variety of ways of examining the integration of cues are being pursued. New information is being obtained so fast that it is difficult for those outside this field to keep up with it. Yet, despite all the exciting new information and all the impressive progress that has been made, we still cannot put together all the known elements to construct orientational and navigational systems that can do what the birds themselves can do. Clearly, there is much more to be learned. The search for the solution to the mystery of avian orientation and navigation must go on.

References

Able, K. P. (1973). The role of weather variables and flight direction in determining the magnitude of nocturnal bird migration. *Ecology*, **54**, 1031–41.

Able, K. P. (1974*a*). Environmental influences on the orientation of free-flying nocturnal bird migrants. *Animal Behaviour*, **22**, 224–38.

Able, K. P. (1974*b*). Wind, track, heading and the flight orientation of migrating songbirds. In *The Biological Aspects of the Bird/Aircraft Collision Problem* (ed. S. Gauthreaux), pp. 331–57. South Carolina: Clemson.

Able, K. P. (1978). Field studies of the orientation cue hierarchy of nocturnal songbird migrants. In *Animal Migration, Navigation, and Homing* (ed. K. Schmidt–Koenig & W. T. Keeton), pp. 228–38. Heidelberg: Springer.

Alexander, J. R. & Keeton, W. T. (1974). Clock-shifting effect on initial orientation of pigeons. *Auk*, **91**, 370–4.

Baldaccini, N. E., Benvenuti, S., Fiaschi, V. & Papi, F. (1975). Pigeon navigation: effects of wind deflection at home cage on homing behaviour. *Journal of Comparative Physiology*, **99**, 177–86.

Barlow, J. S. (1964). Inertial navigation as a basis for animal navigation. *Journal of Theoretical Biology*, **6**, 76–117.

Bookman, M. A. (1977). Sensitivity of the homing pigeon to an earth-strength magnetic field. *Nature*, **267**, 340–2.

Delius, J. D. & Emmerton, J. (1978). Sensory mechanisms related to homing in pigeons. In *Animal Migration, Navigation, and Homing* (ed. K. Schmidt–Koenig & W. T. Keeton), pp. 35–41. Heidelberg: Springer.

Delius, J. D., Perchard, R. J. & Emmerton, J. (1976). Polarized light discrimination by pigeons and an electroretinographic correlate. *Journal of Comparative and Physiological Psychology* **90**, 560–71.

Edrich, E. & Keeton, W. T. (1978). Further investigations of the effect of 'flight during clock shift' on pigeon orientation. In *Animal Migration, Navigation, and Homing* (ed. K. Schmidt–Koenig & W. T. Keeton), pp. 184–93. Heidelberg: Springer.

Emlen, S. T. (1967). Migratory orientation of the Indigo Bunting, *Passerina cyanea*. *Auk*, **84**, 309–42, 463–89.

Emlen, S. T. (1969). Bird migration: influence of physiological state upon celestial orientation. *Science*, **165**, 716–18.

Emlen, S. T. (1970). Celestial rotation: its importance in the development of migratory orientation. *Science* **170**, 1198–201.

Emlen, S. T. (1972). The ontogenetic development of orientation capabilities. In *Animal Orientation and Navigation*, NASA SP-262 (ed. S. R. Galler, K. Schmidt–Koenig, G. J. Jacobs & R. E. Belleville), pp. 191–210. Washington.

Emlen, S. T. (1975*a*). Migration: orientation and navigation. In *Avian Biology*, vol. **5** (ed. D. S. Farner & J. R. King), pp. 129–219. New York: Academic Press.

Emlen, S. T. (1975*b*). The stellar-orientation system of a migratory bird. *Scientific American* 233(2), 102–11.

Emlen, S. T. & Demong, N. J. (1978). Orientation strategies used by free-flying bird migrants: a radar-tracking study. In *Animal Migration, Navigation, and Homing* (ed. K. Schmidt–Koenig & W. T. Keeton), pp. 283–93. Heidelberg: Springer.

Emlen, S. T., Demong, N., Wiltschko, W., Wiltschko, R. & Bergman, S. (1976). Magnetic direction finding: evidence for its use in migratory Indigo Buntings. *Science*, **193**, 505–8.

Gauthreaux, S. A. & Able, K. P. (1970). Wind and the direction of nocturnal songbird migration. *Nature*, **228**, 476–7.

Graue, L. C. (1963). The effect of phase shifts in the day-night cycle on pigeon homing at distances of less than one mile. *Ohio Journal of Science*, **63**, 214–17.

Griffin, D. R. (1969). The physiology and geophysics of bird migration. *Quarterly Review of Biology*, **4**, 255–76.

Griffin, D. R. (1972). Nocturnal bird migration in opaque clouds. In *Animal Orientation and Navigation*, NASA SP-262 (ed. S. R. Galler, K. Schmidt–Koenig, G. J. Jacobs & R. E. Belleville), pp. 169–88. Washington.

Griffin, D. R. (1973). Oriented bird migration in or between opaque cloud layers. *Proceedings of the American Philosophical Society*, **11**, 117–41.

Hartwick, R. F., Kiepenheuer, J. & Schmidt–Koenig, K. (1978). Further experiments on the olfactory hypothesis of pigeon navigation. In *Animal Migration, Navigation, and Homing* (ed. K. Schmidt–Koenig & W. T. Keeton), pp. 107–18. Heidelberg: Springer.

Hermayer, K. L. & Keeton, W. T. (1979). Homing behavior of pigeons subjected to bilateral olfactory nerve section. *Monitore Zoologico Italiano* (N.S.), **13**, 303–13.

Hoffmann, K. (1954). Versuche zu der in Richtungsfinden der Vögel enthaltenen Zeitschätzung. *Zeitschrift für Tierpsychologie*, **11**, 453–75.

Keeton, W. T. (1969). Orientation by pigeons. Is the sun necessary? *Science*, **165**, 922–8.

Keeton, W. T. (1971). Magnets interfere with pigeon homing. *Proceedings of the National Academy of Sciences of the U.S.A.*, **68**, 102–6.

Keeton, W. T. (1972). Effects of magnets on pigeon homing. In *Animal Orientation and Navigation*, NASA SP-262 (ed. S. R. Galler, K. Schmidt–Koenig, G. J. Jacobs & R. E. Belleville), pp. 579–94. Washington.

Keeton, W. T. (1973). Release-site bias as a possible guide to the 'map' component in pigeon homing. *Journal of Comparative Physiology*, **86**, 1–16.

Keeton, W. T. (1974*a*). The orientational and navigational basis of homing in birds. *Advances in the Study of Behaviour*, **5**, 47–132.

Keeton, W. T. (1974*b*). Pigeon homing: no influence of outward-journey detours on initial orientation. *Monitore Zoologico Italiano* (N.S.), **8**, 227–34.

Keeton, W. T. & Brown, A. I. (1976). Homing behavior of pigeons not disturbed by application of an olfactory stimulus. *Journal of Comparative Physiology,* **105,** 259–66.

Keeton, W. T. & Gobert, A. (1970). Orientation by untrained pigeons requires the sun. *Proceedings of the National Academy of Sciences of the U.S.A.,* **65,** 853–6.

Keeton, W. T., Larkin, T. S. & Windsor, D. M. (1974). Normal fluctuations in the earth's magnetic field influence pigeon orientation. *Journal of Comparative Physiology,* **95,** 95–103.

Keeton, W. T., Kreithen, M. L. & Hermayer, K. L. (1977). Orientation by pigeons deprived of olfaction by nasal tubes. *Journal of Comparative Physiology,* **114,** 289–99.

Kiepenheuer, J. (1978). Inversion of the magnetic field during transport: its influence on the homing behaviour of pigeons. In *Animal Migration, Navigation, and Homing* (ed. K. Schmidt–Koenig & W. T. Keeton), pp. 135–42. Heidelberg: Springer.

Kramer, G. (1949). Über Richtungstendenzen bei der nachtlichen Zugunruhe gekäfigten Vögel. In *Ornithologie als biologische Wissenschaft* (ed. E. Mayr & E. Schüz), pp. 269–83. Heidelberg: Springer.

Kramer, G. (1951). Eine neue Methode zur Erforschung der Zugorientierung und die bisher damit erzielten Ergebnisse. *Proceedings of the 10th Ornithological Congress,* pp. 271–80.

Kramer, G. (1952). Experiments on bird orientation. *Ibis,* **94,** 265–85.

Kramer, G. (1953a). Die Sonnenorientierung der Vögel. *Verhandlungen der Deutschen zoologische Gesellschaft.* Freiburg 1952, 72–84.

Kramer, G. (1953b). Wird die sonnenhöhe bei der Heimfindeorientierung verwertet? *Journal of Ornithology,* **95,** 343–7.

Kreithen, M. L. & Eisner, T. (1978). Detection of ultraviolet light by the homing pigeon. *Nature,* **272,** 347–8.

Kreithen, M. L. & Keeton, W. T. (1974a). Detection of changes in atmospheric pressure by the homing pigeon, *Columba livia. Journal of Comparative Physiology,* **89,** 73–82.

Kreithen, M. L. & Keeton, W. T. (1974b). Detection of polarized light by the homing pigeon, *Columba livia. Journal of Comparative Physiology,* **89,** 83–92.

Kreithen, M. L. & Quine, D. (1979). Infrasound detection by the homing pigeon: a behavioral audiogram. *Journal of Comparative Physiology,* **129,** 1–4.

Larkin, T. S. & Keeton, W. T. (1976). Bar magnets mask the effect of normal magnetic disturbances on pigeon orientation. *Journal of Comparative Physiology,* **110,** 227–31.

Larkin, T. S. & Keeton, W. T. (1978). An apparent lunar rhythm in the day-to-day variations in the initial bearings of homing pigeons. In *Animal Migration, Navigation, and Homing* (ed. K. Schmidt–Koenig & W. T. Keeton), pp. 92–106. Heidelberg: Springer.

Lindauer, M. & Martin, H. (1968). Die Schwereorientierung der Bienen unter dem Einfluss des Erdmagnetfeldes. *Zeitschrift für vergleichende Physiologie* **60,** 219–43.

Martin, D. D. & Meier, A. H. (1973). Temporal synergism of corticosterone and prolactin in regulating orientation in the migratory white-throated sparrow (*Zonotrichia albicollis*). *Condor,* **75,** 369–74.

Martin, H. & Lindauer, M. (1977). Der Einfluss der Erdmagnetfeldes auf die Schwereorientierung der Honigbiene (*Apis mellifica*). *Journal of Comparative Physiology,* **122,** 145–87.

Matthews, G. V. T. (1953). Sun navigation in homing pigeons. *Journal of Experimental Biology*, **30**, 243–67.

Matthews, G. V. T. (1955). *Bird Navigation*. London: Cambridge University Press.

Merkel, F. W., Fromme, H. G. & Wiltschko, W. (1964). Nichtvisuelles Orientierungsvermögen bei nächtlich zugunruhigen Rotkehlchen. *Vogelwarte*, **22**, 168–73.

Merkel, F. W. & Wiltschko, W. (1965). Magnetismus und Richtungsfinden zugunruhiger Rotkehlchen (*Erithacus rubecula*). *Vogelwarte*, **23**, 71–7.

Moore, B. R. (1980). Is the homing pigeon's map geomagnetic? *Nature*, **285**, 69–70.

Moore, F. R. (1977). Geomagnetic disturbance and the orientation of nocturnally migrating birds. *Science*, **196**, 682–4.

Nisbet, I. C. T. & Drury, W. H. (1967). Orientation of spring migrants studied by radar. Bird Banding, **38**, 173–86.

Papi, F. (1976). The olfactory navigation system of the homing pigeon. *Verhandlungen der Deutschen zoologische Gesellschaft*, **1976**, 184–205.

Papi, F., Fiore, L., Fiaschi, V. & Benvenuti, S. (1972). Olfaction and homing in pigeons. *Monitore Zoologico Italiano* (N.S.) **6**, 85–95.

Papi, F., Ioale, P., Fiaschi, V., Benvenuti, S. & Baldaccini, N. E. (1978a). Pigeon homing: cues detected during the outward journey influence initial orientation. In *Animal Migration, Navigation, and Homing* (ed. K. Schmidt–Koenig & W. T. Keeton), pp. 65–77. Heidelberg: Springer.

Papi, F., Keeton, W. T., Brown, A. I. & Benvenuti, S. (1978b). Do American and Italian pigeons rely on different homing mechanisms? *Journal of Comparative Physiology*, **128**, 303–17.

Presti, D. & Pettigrew, J. D. (1980). Ferromagnetic coupling to muscle receptors as a basis for geomagnetic field sensitivity in animals. *Nature* **285**, 99–101.

Quine, D. (1979). *Infrasound Detection and Frequency Discrimination by the Homing Pigeon*. PhD thesis, Cornell University.

Richardson, W. J. (1978). Timing and amount of bird migration in relation to weather: a review. *Oikos*, **30**, 224–72.

Sauer, E. G. F. (1957). Die Sternenorientierung nächtlich ziehender Grasmücken (*Sylvia atricapilla, borin* und *curruca*). *Zeitschrift für Tierpsychologie*, **14**, 29–70.

Schlichte, H. J. (1973). Untersuchungen über die Bedeutung optischer Parameter für das Heimkehrverhalten der Brieftaube. *Zeitschrift für Tierpsychologie*, **32**, 257–80.

Schmidt–Koenig, K. (1958). Experimentelle Einflussnahme auf die 24-Stunden-Periodik bei Brieftauben und deren Auswirkungen unter besonderer Berücksichtigung des Heimfindevermögens. *Zeitschrift für Tierpsychologie*, **15**, 301–31.

Schmidt–Koenig, K. (1960). Internal clocks and homing. *Cold Spring Harbor Symposia on Quantitative Biology*, **25**, 389–93.

Schmidt–Koenig, K. & Phillips, J. B. (1978). Local anesthesia of the olfactory membrane and homing in pigeons. In *Animal Migration, Navigation, and Homing* (ed. K. Schmidt–Koenig & W. T. Keeton), pp. 119–24. Heidelberg: Springer.

Schmidt–Koenig, K. & Schlichte, H. J. (1972). Homing in pigeons with reduced vision. *Proceedings of the National Academy of Sciences of the U.S.A.*, **69**, 2446–7.

Southern, W. E. (1972). Influence of disturbances in the earth's magnetic field on ring-billed gull orientation. *Condor*, **74**, 102–5.

Southern, W. E. (1978). Orientation responses of ring-billed gull chicks: a re-evaluation. In *Animal Migration, Navigation, and Homing* (ed. K. Schmidt–Koenig & W. T. Keeton), pp. 311–17. Heidelberg: Springer.

Steidinger, P. (1968). Radarbeobachtungeon über die Richtung und deren Streuung beim nächtlichen Vogelzug im Schweizerischen Mittelland. *Ornithologische Beobachten*, **65**, 197–226.

Wagner, G. (1976). Das Orientierungsverhalten von Brieftauben im erdmagnetisch gestörten Gebiete des Chasseral. *Revue Suisse de Zoologie*, **83**, 883–90.

Walcott, C. (1977). Magnetic fields and the orientation of homing pigeons under sun. *Journal of Experimental Biology*, **70**, 105–23.

Walcott, C. (1978). Anomalies in the earth's magnetic field increase the scatter of pigeons' vanishing bearings. In *Animal Migration, Navigation, and Homing* (ed. K. Schmidt–Koenig & W. T. Keeton), pp. 143–51. Heidelberg: Springer.

Walcott, C., Gould, J. L. & Kirschvink, J. L. (1979). Pigeons have magnets. *Science*, **205**, 1027–9.

Walcott, C. & Green, R. P. (1974). Orientation of homing pigeons altered by a change in the direction of an applied magnetic field. *Science*, **184**, 180–2.

Waldvogel, J. A., Benvenuti, S., Keeton, W. T. & Papi, F. (1978). Homing pigeon orientation influenced by deflected winds at home loft. *Journal of Comparative Physiology*, **128**, 297–301.

Wallraff, H. G. (1972). Nichtvisuelle Orientierung zugunruhiger Rotkehlchen (*Erithacus rubecula*). *Zeitschrift für Tierpsychologie*, **30**, 374–82.

Wallraff, H. G. (1978). Preferred compass directions in initial orientation of homing pigeons. In *Animal Migration, Navigation, and Homing* (ed. K. Schmidt–Koenig & W. T. Keeton), pp. 171–83. Heidelberg: Springer.

Wehner, R. & Labhart, T. (1970). Perception of the geomagnetic field in the fly *Drosophila melanogaster*. *Experientia*, **26**, 967–8.

Williams, T. C., Williams, J. M., Teal, J. M. & Kanwisher, J. W. (1972). Tracking radar studies of bird migration. In *Animal Orientation and Navigation*, NASA SP-262 (ed. S. R. Galler, K. Schmidt–Koenig, G. J. Jacobs & R. E. Belleville), pp. 115–28. Washington.

Wiltschko, R., Wiltschko, W. & Keeton, W. T. (1978). Effect of outward journey in an altered magnetic field on the orientation of young homing pigeons. In *Animal Migration, Navigation, and Homing* (ed. K. Schmidt–Koenig & W. T. Keeton), pp. 152–61. Heidelberg: Springer.

Wiltschko, W. (1968). Über den Einfluss statischer Magnetfelder auf die Zugorientierung der Rotkehlchen (*Erithacus rubecula*). *Zeitschrift für Tierpsychologie*, **25**, 537–58.

Wiltschko, W. (1972). The influence of magnetic total intensity and inclination on directions preferred by migrating European robins (*Erithacus rubecula*). In *Animal Orientation and Navigation*, NASA SP-262 (ed. S. R. Galler, K. Schmidt–Koenig, G. J. Jacobs & R. E. Belleville), pp. 569–78. Washington.

Wiltschko, W. & Wiltschko, R. (1972). Magnetic compass of European robins. *Science*, **176**, 62–4.

Wiltschko, W. & Wiltschko, R. (1975). The interaction of stars and magnetic field in the orientation system of night-migrating birds. *Zeitschrift für Tierpsychologie*, **37**, 337–55; **39**, 265–82.

Wiltschko, W. & Wiltschko, R. (1976). Interrelation of magnetic compass and star orientation in night-migrating birds. *Journal of Comparative Physiology,* **109,** 91–9.

Wiltschko, W., Wiltschko, R. & Keeton, W. T. (1976). Effects of a 'permanent' clock-shift on the orientation of young homing pigeons. *Behavioural Ecology and Sociology* **1,** 229–43.

Yodlowski, M. L., Kreithen, M. L. & Keeton, W. T. (1977). Detection of atmospheric infrasound by homing pigeons. *Nature,* **265,** 725–6.

C.H.LOCKYER and S.G.BROWN

The migration of whales

Introduction

Whales, dolphins and porpoises, the order Cetacea, form a group of highly specialized mammals living an entirely aquatic existence. The majority of cetaceans are marine, and unlike seals and sea otters, they never come on land for any purpose. Strandings are accidental, and usually result in death for the animal. Like land mammals, however, the Cetacea have well defined patterns of feeding, breeding and ecological behaviour and migration is an important factor in the lives of some species.

Migration is here defined as a movement from one place to another taking place periodically. In many cases this periodicity is seasonal and involves regular outward and return movements on a geographical scale.

The order Cetacea includes two suborders of living whales, the Mysticeti (baleen whales) and the Odontoceti (toothed whales). The mysticetes have no teeth. Instead they carry a few hundred baleen plates in their mouths. These plates have on their inner sides bristles resembling coarse hair and these form a dense mat which acts as a highly efficient filter for sifting food from seawater. The baleen whales are notable migrants, most species undertaking highly seasonal return migrations over long distances. The toothed whales also include some migratory species which undertake seasonal migrations but there is apparently no common migration pattern comparable to that found in baleen whales.

The main ecological difference between baleen and toothed cetaceans is in their feeding habits. The baleen whales feed preferentially on planktonic Crustacea, usually copepods, euphausiids and amphipods, which they filter from the seawater. The Greenland right or bowhead whale (*Balaena mysticetus*) and black right whales (*Eubalaena glacialis, E. australis*) which have a fine baleen mesh, feed mainly on the copepods and generally take them in by swimming along and skimming and filtering the seawater continuously. The rorqual whales, blue (*Balaenoptera musculus*), fin (*B. physalus*), sei (*B. borealis*), humpback (*Megaptera novaeangliae*), Bryde's whale (*B. edeni/brydei*) and minke (*B. acutorostrata*) have a coarser baleen mesh and feed on densely swarming plankton. Fin, humpback and Bryde's whales will even take shoaling fish. Their method is to take a gulp and filter and swallow. However, the sei whale may also sometimes swim along skimming and filtering. The gray whale (*Eschrichtius robustus*) sieves bottom mud by scooping and sifting it through its baleen plates.

The dolphins and porpoises feed on fish which they grasp and bite with their teeth. Some of the small odontocete whales like pilot whales (*Globicephala melaena, G. macrorhynchus*) take both fish and squid. Large odontocetes like the sperm whale (*Physeter macrocephalus*) feed almost exclusively on squid, many of benthic origin, although fish may be taken in specific localities.

The feeding of the baleen whales is largely seasonal and in fixed places, whilst that of toothed whales is more widely distributed and probably occurs year round at the same intensity. Food availability is undoubtedly the major factor around which migration is centred.

All whales tend to have a fixed breeding season, but this is especially so in baleen whales. Their mating and calving occur about a year apart in the same areas in low latitudes in winter in contrast to the main summer feeding season in polar waters. Suckling lasts about seven months, taking the calf from winter to the summer feeding grounds where it is weaned. Because of this the reproductive cycle is usually biannual but it may even be annual. The cycle of many dolphins is similar to baleen whales with gestation lasting nine months to a year. In the large toothed whales however, gestation is usually longer than a year. It is 15 months in the sperm whale, and here suckling may last two years or even longer, resulting in a four- or even five-year reproductive cycle.

Most species, if they are at all migratory, tend to give birth in relatively warm waters. However, exceptions are the arctic species, the bowhead, narwhal (*Monodon monoceros*) and white whale (*Delphinapterus leucas*) which have a restricted polar distribution.

Many species of large whales have a cosmopolitan distribution, although northern and southern hemisphere stocks appear to remain discrete. Migration in some of the baleen whales may involve journeys of several thousands of miles annually. In contrast, the Ganges susu (*Platanista gangetica*), one of the river dolphins, may move only a few miles between the main rivers and their tributaries.

Sources of evidence for migration

The evidence for whale migration has come in several ways; some of these have been by direct observation, others more indirectly.

Direct observation

Seasonal density changes reflected in catching operations For species of large whales of commercial importance, catch statistics in the form of logbook records for nineteenth-century whaling vessels, or the detailed and accurate international catch records by species, sex, length and date of capture for twentieth-century whaling published by the Bureau of International Whaling Statistics (BIWS) in Norway, are an important source of information. The economics of whaling operations ensure that the catch statistics are generally a good indication of whale density so long as there are no biases or taboos on catching particular species or categories of whales, and provided the amount of effort directed into catching is known.

Because of the seasonal availability of baleen whales, the antarctic fishery, both pelagic and coastal (South Georgia), operated in summer. The longest whaling season rarely extended beyond October to early April, although the peak months were December to March.

Coastal whaling for baleen whale species in the Southern Hemisphere temperate and tropical waters e.g. off Australia and South Africa, operated only during winter months from about May to September. The fact that the fisheries in different latitudes flourished at different times of the year suggested that their quarry migrated between the two areas. This was especially apparent when over-exploitation of the stocks by pelagic antarctic operations reduced the catches of the coastal whaling stations.

In this situation the species are being caught at or near both ends of their migrations. Many whale fisheries however, only tap one end or part of the migration route. This is especially true in the northern hemisphere where small coastal operations and aboriginal activities are more common. For example, gray whales are hunted by Eskimo groups in the arctic, although they were hunted elsewhere on their migration routes in the past.

Sightings Some cetaceans pass close inshore whilst on migration when they can be fairly easily observed and counted. The gray whale is a prime example of a coastal migrant. In the northeast Pacific Ocean the gray whale migrates annually from the lagoons of Baja California (Mexico) to the Bering Sea and Arctic Ocean north of Alaska and back again. Most of the migratory route hugs the coast. There is a strong directionality in the whales' movements at any time, swimming northward in spring and southward in autumn.

The gray whale has been closely observed in the breeding lagoons at the southern end of its migration where it mates and calves, and also at points on the coast along its migration route.

The duration of the migratory wave past an observation point may be nearly two months, although the peak may only extend over a three-week period (Rice & Wolman, 1971). Other species which hug the coast e.g. the white whale, have been similarly observed in migration.

Species which occur further off-shore are often sighted by ships and low-flying aircraft, when their size, numbers and directions of swimming can be assessed (Bannister & Gambell, 1965). Aircraft are particularly good for making observations, since disturbance of the animals is not usually so likely to occur as when vessels are used. However, ships are used for oceanic searching because of their longer range.

For density estimates from such surveys, the number of miles steamed or flown and the range of visibility must be accurately known, and possible duplicate sightings must be avoided. Fig. 1 illustrates the estimated variation in numbers of large baleen whales in ice-free antarctic waters derived from sightings from a research vessel. The peak abundance occurs in summer and very few whales are present in winter. The latter animals which are possibly juveniles, may either get 'out of step' with the main migratory movements or simply not migrate over one or more years.

Recognition of individuals by means of artificial and natural markings Catch statistics and sightings surveys provide information about the movements of a population as a whole. The movements of individuals can be followed by marking experiments or by the recognition of animals bearing distinctive natural markings.

The 'Discovery' whale mark has been extensively used for marking large whales since the 1930s (Brown, 1977a). It consists of a stainless steel tube approximately 23 cm long fitted with a leaden ballistic head and bearing a serial number and address for return. The mark is an internal one. It is fired from a

Fig. 1. Estimated seasonal variation in the numbers of large baleen whales (blue, fin and humpback) present in ice-free antarctic waters. Dotted line, mean; solid line, best estimate. (After Mackintosh, N. A. & Brown, S. G. (1956). *Norsk Hvalfangst-Tidende* **45**(9), 469–80.)

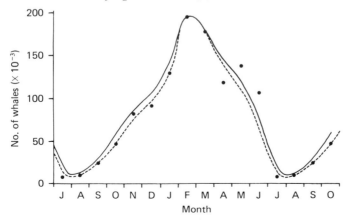

modified 12-bore gun into the back of the whale at sea and is designed to remain buried in the muscles of the back and be found when the animal is later killed and processed at a whaling factory. A smaller version fired from a modified .410 gun has proved suitable for marking calves and smaller species, e.g. minke whales. A reward is paid for marks returned with information on the date and position of capture of the whale, its species, sex, length, and other data if available. The mark only provides data on two points in time, however, and no details about movements between these points unless the whale is marked again before capture. Since recoveries are dependent upon whaling, the usefulness of the 'Discovery' mark grows less with the decline of the industry. Recently attempts have been made to redevelop the mark to include an externally visible streamer which permits short-term visual tracking (Brown, 1978).

Small cetaceans have been marked in a variety of ways. A button tag consisting of two circular discs pinned together through the dorsal fin has been used. The discs bear a number visible through field binoculars. Spaghetti tags (more usually used on fish), freeze branding and waterproof luminescent paint have all been used. All these are visual marks and can be used independently of a fishery (Brown, 1978).

Permanent distinctive natural markings have recently been used in the recognition of individual whales. Right whales off Patagonia have been successfully identified from season to season by individual characteristic patterns of the callosities on the 'bonnet' or rostrum (Payne, 1976). Aerial photographs provide a form of identikit of each whale. Tail fluke profile and pigmentation of humpbacks have proved successful in identifying individuals (Katona et al., 1979). The individual shape of dorsal fin profiles has been used in dolphins (Würsig & Würsig, 1977). Some animals have even more specific identifying marks and scars (see Fig. 2) which enable the individual to be recognized instantly wherever it goes over numbers of years (Lockyer, 1978a).

Indirect observation

External parasites and commensals Some external parasites only flourish or are acquired in particular environmental conditions. *Penella balaenopterae*, a sessile crustacean, takes root in whale blubber and trails like a wire from the body. It is more frequently found on whales in the tropics and warm waters than in polar seas where it tends to die and fall off (Slijper, 1962). As the infection originates in warm waters, animals found with this parasite in arctic and antarctic waters are clearly recent visitors from lower latitudes.

A commensal acquired in polar waters which tends to die off in warm waters is the diatom, *Cocconeis ceticola* which proliferates and forms a brownish film on the whale's skin (Hart, 1935; Omura, 1950). Visitors to the antarctic from low latitudes are usually quite clean but gradually acquire a diatom film about one month after arrival. Conversely, new arrivals from the antarctic to warmer waters can often be identified by remnants of skin diatom film which soon disappears. Fig. 3 illustrates this seasonal migration as detected by presence or absence of diatom film.

General pattern of migration in mysticetes

The migratory species include in both hemispheres the blue, fin, humpback, minke, sei and black right whales. There is direct evidence of migrations from mark returns for fin, sei and humpback. There is also some evidence that Bryde's whales migrate, although perhaps not as extensively as other rorquals, being at all times a warmer-water resident.

The gray whale population in the northeast Pacific Ocean migrates extensively. This is the only place where it survives in any numbers. The present status of the population in the northwest Pacific off the coasts of Korea and Japan, where it formerly occurred in some numbers, is uncertain, and the species no longer exists in the North Atlantic Ocean.

The bowhead whale is exclusively arctic and never migrates outside the polar zone. In the Southern Hemisphere the pigmy blue whale (*Balaenoptera musculus*

Fig. 2. Dolphin of the species *Tursiops truncatus* with clear permanent scars on the head: white scar near the blowhole; and deep pit above the right eye.

brevicauda) is probably migratory in the same way as other rorquals, but information on its general distribution and seasonal occurrence is limited.

The form of migrations in the blue, fin, humpback, minke, gray and right whales, and to a certain extent the sei can all be generally described as following a similar pattern. This pattern is a seasonal migration between subtropical to temperate waters for breeding in winter and polar or subpolar waters for intensive feeding during summer. The whole cycle of migration is repeated annually. In Fig. 4 the pattern of migration is shown schematically. The same seasonal links

Fig. 3. Incidence of diatom infection in relation to movements of Southern Hemisphere blue whales.

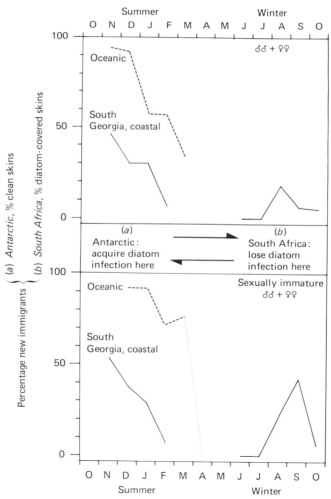

with breeding and feeding take place in both hemispheres although of course Northern and Southern Hemispheres are six months out of phase, which means that in theory both hemisphere populations move northwards and southwards in unison, so reducing the chances of Northern and Southern Hemisphere stocks ever mingling even if their winter ranges overlap. There is no evidence from mark returns that they cross from one hemisphere to another. However, most of the knowledge of the geographical range of migration has been obtained from catch records. The distances covered are of the order of several thousand miles between latitudinal limits of 15° and 70° in species like the humpback. Longitudinal movements in fin whales in antarctic waters are generally limited (Brown, 1954) which suggests that discrete stocks of this species may occur within certain longitudinal boundaries especially where reinforced by natural geographical boundaries such as land masses. This is probably also true for other species. Some may form up to six discrete stocks in the antarctic (Hjort, Lie & Ruud, 1932; Mackintosh, 1942; Omura, 1973; Budylenko, 1978; Doroshenko, 1979).

Fig. 4. General form of whale migrations. (After Mackintosh, N. A. (1965). *The Stocks of Whales,* 232 pp. London: Fishing News (Books) Ltd.)

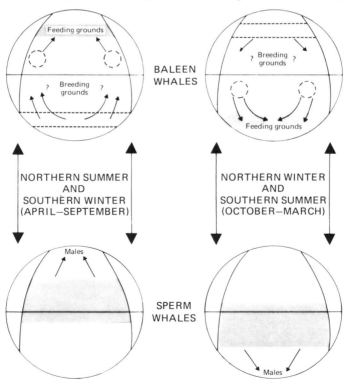

Most of the detailed information is known for the species whose established migration routes pass close in to land where they have been observed over numbers of years. These species are humpback whales in the Southern Hemisphere and gray whales. Although much is known about blue and fin, especially on the polar feeding grounds, their breeding grounds and routes of migration are still unknown. Slightly more can be said of the southern black right whale which comes into sheltered coves on the southern coasts of South America, South Africa, Australia and New Zealand where it has been observed in some detail in Patagonia and around the Cape Province, South Africa.

Humpback whale migration in the Southern Hemisphere

There are at least six stocks of humpback whales in the Southern Hemisphere. Their breeding grounds are on the east and west coasts of South America, Africa, Australia and among the islands of the South West Pacific Ocean. Southward migration occurs in spring to the antarctic where the stocks concentrate in five areas. These are in the South-East Pacific, South Atlantic, south of South Africa, South-East Indian Ocean and south of New Zealand (Dawbin, 1966). Mark recoveries suggest that there is overlap of the stocks on the feeding grounds in the antarctic, and possibly a small exchange between the stocks on the breeding grounds themselves. Fig. 5 illustrates the distribution and movements of the southern humpback stocks. The breeding grounds of the humpback are warm-water (25°C) coastal regions in the tropics. The females mainly conceive in August and also give birth here about eleven and a half months later. The calves are apparently suckled for up to eleven months before weaning takes place again in the winter tropical waters.

The southward spring migration is led by recently pregnant females and immature males, followed by resting females and mature males, and finally females in early lactation with their calves. Whilst on the grounds, the various classes appear to mix randomly. The return autumn migration is led by the lactating cows and their calves, followed by a succession of whales in approximately the reverse order of arrival.

The pregnant cows may spend as long as six and a half months from late November to May in waters south of 60° S. Most whales stay about five and a half months, but the lactating females stay only four and a half months.

Gray whale migration in the northeast Pacific Ocean

The breeding grounds are generally shallow sheltered lagoons around Baja California (Mexico) (Gilmore, 1960). The whales are there between late December and early March. The waters range between warm temperate to tropical.

The majority of gray whales spend the summer months from late May through

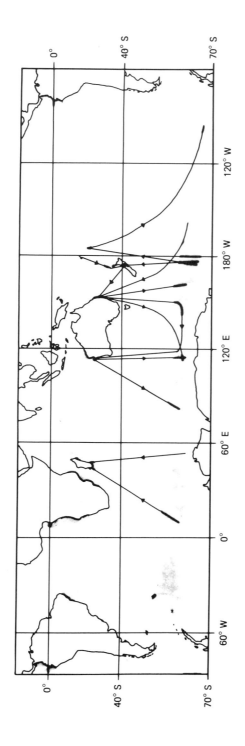

October in the shallow waters of northern and western Bering Sea, Chukchi Sea and western Beaufort Sea in the arctic (Rice & Wolman, 1971). The northern-most boundary of the summer distribution seems to be the close pack ice which may hinder their movements. Gray whales not infrequently occur in broken pack ice. In all places, the gray whale tends to remain in shallow continental shelf areas mainly because of its feeding habits.

During migrations the gray whale swims parallel to the coastline only a few kilometres off shore when there are headlands and in places where the continental shelf is narrow. However, in regions where there are deep coastal indentations, the gray whale usually cuts across up to 200 km off shore to take the most direct route. This had been observed (Rice, 1965) in the area of Point Conception, at latitude 34°27′ N, where the gray whale moves through the Channel Islands. A similar crossing occurs from Punta Baja, latitude 29°57′ N, to Isla Cedros, latitude 28°22′ N, in Baja California.

The migration, both south and north, is usually led by adult females. The males follow with the immature whales. The newly pregnant cows are the first to move northwards at the end of the winter breeding season. Rice & Wolman (1971) give details of the exact peak timings of migration of each sexual class. Generally the whales spend between two and three months on the winter breeding grounds. Females which have newly conceived stay the shortest time and *postpartum* females the longest time.

Factors influencing migration in mysticetes

Food availability

Food, feeding and fattening This is the major factor influencing mysticete migration. The polar seas are highly productive compared with temperate and tropical seas. In the antarctic, the major food of baleen whales is *Euphausia superba,* a swarming planktonic crustacean (Marr, 1962). The distribution of this crustacean, known commonly as krill, is circumpolar and extends from the antarctic shelf waters in some sectors up to the antarctic convergence (Deacon, 1937; Marr, .1956, 1962; Mackintosh, 1972). The most productive areas of the polar waters are between 60° W and 60° E (Mackintosh, 1973), encompassing the Weddell Sea and parts of the South Atlantic and South West Indian Oceans. In Fig. 6 the distribution of the post-larval krill from net catches is shown in relation to the major areas of baleen whaling. The two distributions coincide very closely indicating strong links between predator and prey. Baleen whales may consume

Fig. 5. Humpback whaling grounds and migration routes. Shaded areas, old and modern whaling grounds; arrowed lines, migration routes as shown by captures of marked whales.

other locally swarming plankton e.g. *Euphausia crystallorophias* in the Ross Sea (Marr, 1956).

In the Northern Hemisphere, diet is rather similar to that in the Southern Hemisphere. Baleen whales take mainly swarming euphausiids. However, shoaling fish are more readily taken e.g. saury (*Cololabis saira*), herring, Alaska pollack and mackerel in the Pacific regions (Nemoto, 1959) and capelin, anchovy, etc. in the Atlantic. The sei whales consume *Calanus plumchrus* and *C. cristatus*, both copepods, in the Bering and Chukchi Seas (Nemoto, 1963); squids and other items are probably taken incidentally.

Fig. 6. Antarctic whaling grounds and distribution of krill. Shaded areas, distribution of pre- and post-war pelagic whaling, all year; dotted lines, outermost limit of moderate concentrations of krill (inner limit bounds zone of low concentration). (After Mackintosh, 1973.)

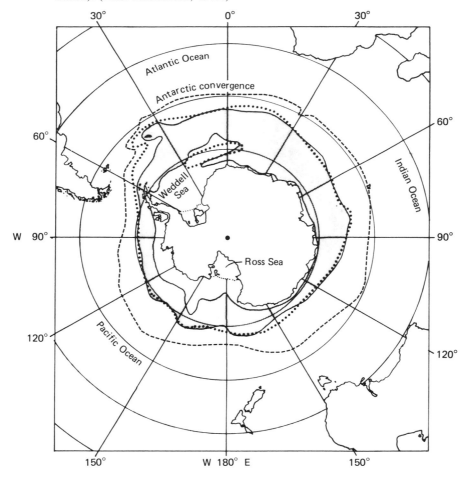

In contrast, food availability outside the main polar and subpolar summer feeding grounds is rather less, and feeding is probably opportunistic. Certainly plankton density is considerably lower, so rendering filter feeding less efficient.

In the antarctic during summer rorquals generally gorge themselves with food. Fig. 7 shows that in fin whales (typical of blue, minke and humpback also), generally between 70 and 100% of whales examined have food in the stomach. In the temperate waters off South Africa in winter, however, less than 60% have any food. In terms of quantity too the amount in winter is only about one fifth to one twentieth of the summer intake (Gambell, 1968*a*). In general therefore, daily winter food consumption probably averages one tenth or less of that in summer, which is generally reckoned to be about 35–40 g of food/kg body weight (Lockyer, 1976).

Because of seasonal limitation on feeding, all migratory mysticetes must store energy during the main summer feeding period, as a reserve for the lean winter months. This they do by storing fat and oil in the blubber which increases measurably in thickness (Rice & Wolman, 1971; Lockyer, 1976). Fat is probably also stored in other tissues such as muscle, viscera and bone as often reflected towards the end of summer in the increased commercial oil yields which are

Fig. 7. Seasonal incidence of feeding in Southern Hemisphere fin whales. Solid line, sexually mature; dashed line, immature; sample sizes in square brackets. Lower left, temperate waters, Durban, South Africa; upper right, polar waters, antarctic.

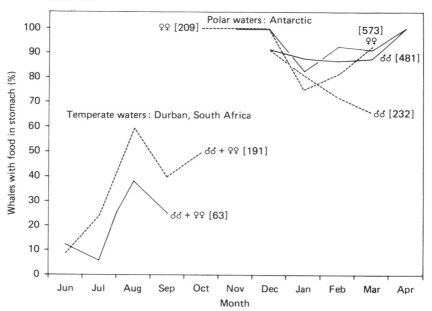

greater than those anticipated from blubber alone (Lockyer, 1976), although not all whale tissue is rendered for oil. Also the total body weight of the whales increases between spring and the end of summer. In Fig. 8 the trend of increase in body weight for Southern Hemisphere blue whales is expressed as a ratio of body tissue to skeletal weights from November to March. The dip between early February and early March is almost certainly a reflection of a secondary influx of lean animals onto the feeding grounds. Ash (1953) has produced similar weight-increase patterns over the summer feeding season for humpback whales in the antarctic. Apparently, baleen whales may increase their body weight by 50 to 100%. All this occurs during a four to six and a half month season on the grounds.

Rice & Wolman (1971) presented calculated data on weight losses from girth measurements for gray whales in the northeast Pacific, and calculated that up to 29% of body weight was lost between the southward (breeding) and northward (feeding) migrations, equivalent to about 40% lean body weight increase after

Fig. 8. Body weight increase in blue whales during the summer feeding season in the antarctic. Solid circles, males + females; open circles, pregnant females; arrow, secondary influx of lean migrants.

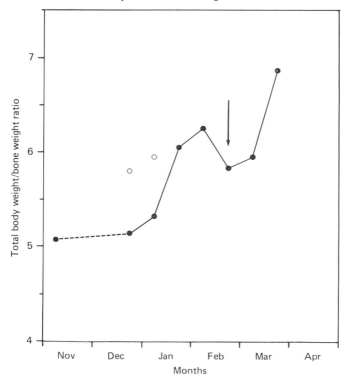

feeding, but energy losses during migration and breeding are probably quite high, at a rate similar to that in hibernating and starving mammals i.e. 0.21–0.37% body weight/day (Kayser, 1961; Lockyer, 1976).

Growth Little is known about the detailed pattern of growth in mysticetes. However, one might anticipate that developmental growth as opposed to temporary fat storage occurs in steps correlated with feeding rather than continuously.

There is indirect evidence that food supply does affect growth rate quite markedly. Southern Hemisphere fin, sei and minke whales have all shown a faster growth rate and consequential earlier sexual maturation, almost certainly due to the massive depletion of major commercially exploited large baleen whale species since the 1930s (Lockyer, 1972, 1974, 1977, 1979*a;* Masaki, 1979). The obvious link with growth is a greater food supply and less competition in feeding.

The growth of the newborn calf is of course independent of food availability except through the mother, for the first seven months or so of life. At this time, when it is weaned, the calf is already on the summer feeding grounds. However, food availability in lower latitudes during winter and autumn is probably important to the calves of minke and humpback whales which suckle their young for about four to eleven months respectively. The growth of the calf thus apparently becomes dependent on a food supply other than milk outside of the main feeding grounds.

Reproduction

Gestation and growth of the foetus The gestation period of all baleen whales is about a year. The gray whale has the longest gestation of just over thirteen months, whilst the other baleen whales carry the foetus for ten to eleven and a half months. Baleen whales are believed to be monogamous. Conception generally takes place in winter, May or June for Southern Hemisphere blue, fin, sei and minke whales, May to July for southern right whales, and July to August for southern humpback whales. The Northern Hemisphere stocks of these species are less well documented, but are probably six months out of phase with their southern relatives. The northeast Pacific gray whale has peak conceptions during December.

The females of the species usually only ovulate at about this time, and the males of some species, e.g. fin and humpback whales, have a seasonal testicular activity associated with winter breeding (Laws, 1961; Chittleborough, 1965). Both sexes appear to come into breeding condition independently of sexual stimulation. This strongly seasonal gonadal activity has a high survival value in that breeding and sexual activities only occur when environmental conditions are appropriate.

The foetal growth curve of rorquals, which commences linearly, has an exponential spurt approximately five months after conception (Laws, 1959), coinciding with the migration to the summer feeding grounds. This latter intense growth phase of the foetus ensures that it reaches term by the time the female has returned to the winter grounds.

Calving and lactation Once returned to the winter grounds, the female gives birth. The breeding and calving grounds of most baleen whales are unknown. However, right whales favour sheltered coves and bays in warm temperate regions for breeding (Donnelly, 1969), and gray whales seek out shallow lagoons along the coast of Baja California (Rice & Wolman, 1971). Like the timing of conception, the birth dates automatically occur within a small range of months.

The cow suckles her calf during the winter months and weans the offspring on the first summer feeding migration, with the two exceptions (minke and humpback) as already noted in the subsection on growth. The cow is extremely thin by the time the calf is weaned, so that subsequent pregnancies must depend for their success on her feeding well during summer.

The whole reproductive cycle in most baleen whales is two years. This comprises about one year of gestation, half a year lactating and half a year resting before recommencing the cycle. This is schematically shown in Fig. 9 for the Southern Hemisphere female fin whale, and relates events in the reproductive cycle to migration and feeding. It can be seen that the pregnant cow remains longer on the polar feeding grounds than the resting female, presumably accumulating energy reserves for lactation. The one possible exception to the two-year reproductive cycle is the minke whale which is currently believed to reproduce annually (International Whaling Commission, 1979).

Energy cost of reproduction The energy cost of reproduction is relatively high being between 25 and 35% more than that of normal growth and maintenance requirements (Lockyer, 1978b). The cost is proportionately greater in the smaller and younger growing whale than the full adult. The cost of pregnancy would appear to be easily absorbed, but lactation is a considerable energy drain being 12–15 times more than that of pregnancy (Lockyer, 1976, 1978b). In Fig. 8 the pregnant females are noticeably fatter, and their blubber thickness is always relatively greater than in other whales (see previous references). This is probably a direct result of a longer period spent feeding during summer.

Since twentieth-century exploitation reduced the southern baleen whale

Fig. 9. Reproductive cycle in Southern Hemisphere female fin whale.

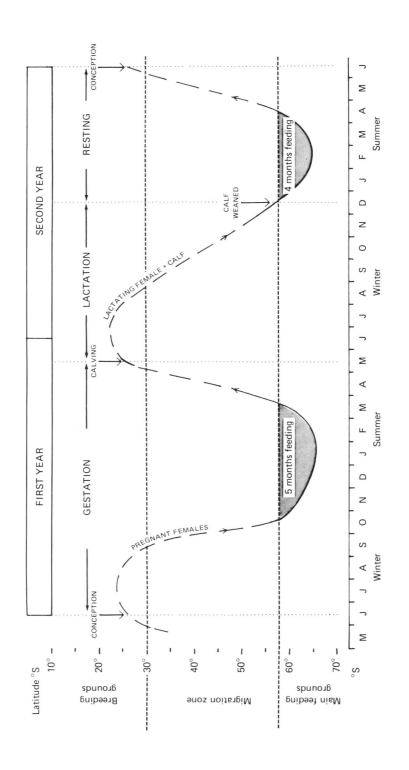

stocks, there has been a statistically significant increase in the numbers of blue, fin and sei whales pregnant each year (Gambell, 1973) from the 1930s onwards. One simple explanation apart from generally improved fecundity, is that many of these whales must be simultaneously lactating and pregnant, i.e. the reproductive cycle may be annual for some individuals. The implications of this are that food availability must be considerably more favourable in recent years, because of less competition on the antarctic feeding grounds.

Competition

There is no direct evidence for inter- or intra-specific competition. The faster growth rates, earlier sexual maturation, and increased proportion of pregnant whales all point to a greater abundance of food supply which over-exploitation of whale stocks would release to the remaining animals. However, if animals of different species all feed simultaneously in the same places on the same food then competition is likely under certain circumstances e.g., when conditions for feeding are generally unfavourable. In this situation, the Southern-Hemisphere stocks of whales may have been in competition in past years (pre-1930), before the stocks were exploited intensively, but are not necessarily so now. Certain species are more likely to be in competition than others because of specific feeding habits e.g. blue and minke (International Whaling Commission, 1979), sei and right (Kawamura, 1978).

Data on a southerly shift in the southern limit of distribution of the Southern Hemisphere sei whales have been produced as evidence of reduced inter-species competition (Nemoto, 1962), but it is not certain whether or not this is due wholly or partly to changes in oceanographic conditions.

Climate and physical environment

In the polar waters, the advance of the ice edge may physically exclude whales from an area. Ice cover in winter also prevents access to the krill. In Fig. 10 the mean outer and inner limits of the antarctic pack ice are shown. In spring, from October onwards the whales, e.g. blue, follow the retreating ice edge in towards the continent. In autumn, from March onwards when the ice starts to advance again, the majority of the rorquals begin to leave antarctic waters. A few whales do get accidentally trapped in the ice (Taylor, 1957), but this is unusual.

The warm waters of up to 25°C in the breeding grounds probably have no advantage to adult whales which have a greater problem staying cool than warm. However, newly born calves have extremely thin blubber which contains little oil or fat as insulation, so that birth in warm waters has significant survival value for them until they can build up their blubber.

It has been suggested that variations in daylength may stimulate migration in whales (Dawbin, 1966). As the spring movement towards the poles occurs, day-length increases dramatically until in midsummer daylight exists for virtually 24 h. With the onset of autumn, the day length shortens rapidly and this may initiate movement back to warmer waters.

Influence on migration and distribution of a different kind is the possible inter-ference in whales' movements by boat traffic suggested by Nishiwaki & Sasao (1977). They reported a decreasing frequency in the representation of young calves in the catches off Japan in the Korean strait over a ten-year period, and believed this to be due to a shift in the migration route caused by traffic distur-bance.

Social organization of migration in mysticetes

The social organization and succession of different social classes in migration have already been outlined for humpback and gray whales. Dawbin

Fig. 10. Seasonal antarctic ice cover (see also Fig. 6 for whale and krill dis-tributions).

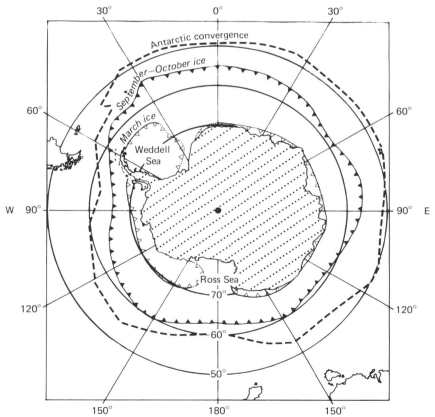

(1966) and Rice & Wolman (1971) give further details. Such organization has been observed for other baleen whales, chiefly from the abundance of certain animals at different times during the whaling season. The migration occurs as a procession or protracted wave of animals rather than a mass movement. Mackintosh (1966) states that arrivals in the antarctic occur between September and January, and departures may extend from March to as late as July. Generally pregnant females arrive on the polar feeding grounds first and remain the longest time, perhaps five and a half months in blue and fin whales. Adult males and resting females follow, with a vanguard of juveniles up to one month later (Lockyer, 1976). The return migration follows a similar order usually after about four months' stay in antarctic waters in blue and fin whales. Lactating females and their calves are the latest arrivals on the polar grounds. In Fig. 8 it was noted that a secondary late influx of migrants showed in the temporary break in the trend of fattening. Early departure of fattened animals could also augment this break.

Apart from this procession of different classes of animals, there is also geographical segregation in varying degrees. Juvenile fin whales tend not to penetrate as far south as the adults. This is evidenced by the low proportion of immature whales in pelagic antarctic catches which occur further south than the coastal catches around South Georgia (Bureau of International Whaling Statistics) where immature animals formed a significant proportion of the catch.

Sei whales segregate latitudinally and catches in the antarctic are virtually all sexually mature adults (Bureau of International Whaling Statistics). Immature whales remain mainly in sub-antarctic waters. Elsewhere the distribution of the sei whale is rather variable. Off Iceland for example, sei whales appear in numbers only irregularly as reflected in the catch records (Bureau of International Whaling Statistics). Sei whale distribution is probably influenced by water temperature and oceanographic conditions here, as in the Southern Hemisphere (Nemoto, 1959, 1962; Kawamura, 1970, 1974).

The minke whale appears to have a marked geographical sexual segregation. Details are lacking but catches both in Northern and Southern Hemispheres show very uneven representation of the sexes in certain areas and months (Ohsumi & Masaki, 1975; Lockyer, 1979b; Christensen, 1979).

Generally speaking all baleen whales have a north/south orientation in migration, but the precise timing of migration and distances covered vary according to the species, sexes and ages of the animals.

Speed and distances in migration

The most extensive migrants are the gray, humpback, blue, fin and minke whales. These species cover more ground in terms of latitude than sei, Bryde's and right whales.

The distance covered annually on a return migration between latitudinal limits in each hemisphere of e.g., 20° and 65°, would be between 5 000 and 6 000 nautical miles. Observations on migrant gray whales suggest that they do not stop at all during migration. The speed at which such a distance can be covered depends on the average sustained swimming speed. Brown (1977*b*) noted a marked sei whale which covered at least 2 200 nautical miles in antarctic waters within 10 days, an average continuous speed of about nine knots.

The observed range of migratory speeds (Lockyer, 1976) for blue, fin, sei, Bryde's and minke whales are all in the range 3–16 knots. The humpback at 2–8 knots, right 3–6 knots, and gray 2–8 knots, are rather slower. Dawbin (1966) has calculated that average humpback migration speed is about 15° of latitude per month, or less than two knots, a result perhaps of short local movements. Apparently there are no differences in speed between the sexes and age classes.

Lockyer (1976) and Kawamura (1975) have calculated that it is theoretically possible for a blue or fin whale to migrate between winter and summer grounds in about a week if a continuous swimming speed of about 15 knots were maintained.

Pike (1962) has observed in the northeast Pacific that gray whales on the southward breeding migration average about 4.2 knots over the entire range. During daylight they swim slightly faster (about 0.75 knots more) than at night time. The northward feeding migration speed is only about half as fast. These observations are based on the time of passing between observation points all along the route.

General pattern of migration in odontocetes

As already noted much of the evidence for migrations in cetaceans is derived directly or indirectly from the records of, or access to, whales provided by whaling. The sperm whale has been hunted since the early eighteenth century and sperm whaling has spread world-wide. However, few other odontocete species have been widely hunted on a large scale, so information about possible migrations in many species is absent. The available evidence suggests that the pattern of regular annual large-scale latitudinal seasonal migrations, intimately linked with the annual reproductive and feeding cycles, which is found in most of the mysticete species is uncommon in odontocetes. The sperm whale, northern bottlenose whale (*Hyperoodon ampullatus*) and white whale can be used as examples of various forms of migration in odontocetes.

Sperm whale

The sperm whale, the largest of the odontocetes, is polygynous. It is the most sexually dimorphic in size of all living cetaceans and has a complex social organization (Best, 1979). The main body of sperm whale populations is found in tropical waters. Charts of the monthly catches of nineteenth-century American sperm-whaling vessels (Townsend, 1935) show that sperm whales are present in tropical and equatorial waters throughout the year. However, in the Northern Hemisphere summer there is a movement towards Northern Hemisphere waters. In the Southern Hemisphere summer there is a movement towards Southern Hemisphere waters. In the winter in each hemisphere there are consequent movements towards the Equator from tropical waters (Fig. 4).

Female sperm whales and small males occur as mixed schools of both sexes and as separate schools of small bachelor males, mainly in tropical and subtropical waters. Best (1979) produces evidence suggesting that in the Southern Hemisphere the southward limit of their movements is related in some way to the position of the subtropical convergence. However, female whales have been found further south in the Southern Ocean. In the North Pacific Ocean females are found up to about 57° N (S. Ohsumi & K. Nasu, unpublished data).

Medium-sized bachelor males and larger males migrate into temperate and polar waters in both hemispheres. Direct evidence for these migrations comes from the results of whale marking in the North Pacific (Omura & Ohsumi, 1964; Ivashin & Rovnin, 1967) and Southern Hemisphere (Best, 1969). In the Southern Hemisphere evidence for the return of male whales from high latitudes to warmer waters is given by the presence of the antarctic diatom *Cocconeis ceticola* on the skin of animals examined in South Africa, showing that they had recently migrated from colder waters (Best, 1969) and by the presence of the beaks of antarctic squids in the stomach contents of whales examined at Durban (Clarke, 1972). It is believed that these migrations of male sperm whales to and from high latitudes are regular annual seasonal return migrations (Best, 1979) but the evidence for this is not as clear as in baleen whales.

There are believed to be separate eastern and western populations (stocks) of sperm whales in the North Pacific Ocean. Evidence from whale marking suggests that the large males from both stocks occur together in the region between 160° E and 160° W north of 50° N. It is assumed that these males return to their respective stocks on their southward migrations. There is little information about possible separate populations in the North Atlantic. In the Southern Hemisphere nine stocks are recognized for management purposes (International Whaling Commission Sperm Whale Divisions 1–9). These are based mainly on the density distribution of catches and sightings and there is little information on the degree of separation between them or the relationship of sperm whale migrations to them.

Northern bottlenose whale

The northern bottlenose whale, one of the Ziphiidae (beaked whales), is widely distributed over deep water in the North Atlantic Ocean. On the west-

ern side it is recorded from 70° N in the Davis Strait as far south as Rhode Island
(41° N), and on the eastern side from 80° N off Spitsbergen as far south as the
Cape Verde Islands (15° N). There are local concentrations in some areas which
may represent separate stocks (Benjaminsen, 1972).

An outline of migrations in this species has been derived mainly from the
distribution of catches by Scottish and Norwegian whalers in the late nineteenth
and early twentieth centuries, and by the modern Norwegian 'small whaling'
fleet; together with records of sightings at sea and of coastal strandings. Little is
known of the wintering areas of the species. Mitchell & Kozicki (1975) suggest
that in the western North Atlantic they lie on the continental slope and farther off
shore from Cape Cod to the Grand Bank (42°–45° N). They are apparently much
farther south on the eastern side of the ocean. The whales are believed to move
northwards towards arctic waters in March. They are found as far north as 60° N
off the Labrador coast in May and June, though there is also a concentration
much farther south off Sable Island (44° N) in June and July. On the eastern side
they occur in the waters between Iceland and Jan Mayen in May and early June,
and also farther north to the west of Spitsbergen. Most of the whales appear to
leave the northernmost waters before the end of June, and Icelandic waters by
early August, but some are found off the Norwegian coast as late as November
(Benjaminsen & Christensen, 1979). Strandings on the British coasts are concentrated from July to October, and on continental North Sea coasts from August to
November, and these are presumably of animals migrating southwards. The wintering areas are assumed to be occupied from December to February.

White whale

The white whale occurs in arctic and sub-arctic coastal and estuarine
waters though occasional specimens have been recorded much further south.
There are a number of separate or apparently separate populations in Alaskan,
Canadian, Greenland and Euroasian waters. The pattern of seasonal distribution
and migration is different in different populations. The populations of Alaskan
and the western Canadian Arctic waters may be taken as examples.

In Alaskan waters the population in the Gulf of Alaska is centered on Cook
Inlet where the animals apparently remain throughout the year. This population
is believed to be isolated from those in the Bering Sea to the north of the Alaska
Peninsula. In the Bering Sea there is a resident population in the Bristol Bay
area, and a migratory population which is believed to winter in the Bering Sea.
Part of this migratory population spends the summer south of the Bering Strait
while the remainder migrates through the strait in March and April. Some move
into eastern Siberian waters while others migrate to western Canadian waters,
passing through the Chukchi Sea in April and crossing the Beaufort Sea in May
and June to Banks Island and Amundsen Gulf and then to the Mackenzie Delta.

The exact route followed each year probably depends upon ice conditions (H. W. Braham, unpublished data). The animals are in the Mackenzie estuary during most of July and leave at the beginning of August. Some may return to the Amundsen Gulf before migrating westwards out of the Beaufort Sea in September to return to the Bering Sea (Fraker, 1980).

Factors influencing migration in odontocetes
Food availability

In mysticete whales the relation between their mainly seasonal feeding and the pattern of their migrations is clear. However the relationship of feeding to migrations is less obvious in odontocetes, which are assumed to feed throughout the year. Baker (1978) suggests that their migrations are based on the seasonal succession and migration circuits of their prey species. Certainly in some areas the movements of some species of odontocetes are closely related to the distribution of their food. For example, Mercer (1975) and Sergeant & Fisher (1957) show that the distribution of pilot whales in Newfoundland waters is closely linked with the distribution of the short-finned squid (*Illex illecebrosus*) on which they feed in this area. The occurrence of this squid appears to be determined mainly by water temperature. In Trinity Bay, Newfoundland, where there was a fishery for pilot whales, the dates of their arrival and departure were very closely linked to the corresponding dates for the squid and catches of both animals in Newfoundland waters are also closely linked.

Sperm whales appear to feed throughout the year. Examination of the stomach contents of sperm whales in South African and antarctic waters gives no evidence of clear seasonal fluctuations in the amount of food taken. There is considerable random variation in the amount of food, however, and this is perhaps dependent on the immediate availability of the squid forming the bulk of the diet. Measurements of blubber thickness also do not show seasonal variations. This supports the conclusion that there is no marked seasonal variation in food consumption and also suggests that food is never scarce enough to require sperm whales to use their blubber as a substantial source of energy (Gambell, 1972). There is as yet no positive evidence of a link between the migration of sperm whales and the distribution of their food. However, Best (1979) suggests that if, as reported, squid are less numerous both in species and in numbers in the antarctic than warmer waters, this might explain why mixed schools of sperm whales and schools of small bachelors are not found in this region, since their numbers would be incompatible with the available food supply. Within the antarctic, Holm & Jonsgård (1959) have shown that sperm whales tend to concentrate at times around banks. This distribution is related to phases of the moon, and presumably is associated with the behaviour of the squid on which they feed. In warmer waters off Durban, South Africa, an association of sperm whales with current lines which may indicate areas richer in food organisms has been noted (Gambell, 1968b).

The northern bottlenose whale feeds on squid, especially the species *Gonatus fabricii*, and on species of deep-water fishes. In summer its distribution over deep water may be related to the distribution of the squid, and Benjaminsen & Christensen (1979) suggest that the autumn strandings on the coasts of northwest Europe could be associated with an inshore movement of the squid at this time. They describe the whales' spring and early summer migration to arctic waters as a feeding migration, but little is known about feeding at other times of the year, and this description may not be correct.

The white whale feeds on a wide range of shallow-water organisms including decapod crustaceans, cephalopods and schooling fish. While feeding occurs in shallow water, little apparently takes place while the whales are in the river estuaries, and they have been observed to feed off shore. White whales appear to be opportunistic feeders, and form concentrations in areas where food is abundant during the summer.

Reproduction

In the species of odontocetes which are recognized as migratory there is as yet little clear evidence of a close link between the length of the gestation period and the migratory cycle such as exists in mysticetes. However, a link with an annual migratory cycle may exist in those species with a gestation of 12 months or less. In the northern bottlenose whale, pregnancy is believed to last for 12 months, with a peak of mating and parturition in April (Benjaminsen, 1972). Lactation may last for 12 months and it is likely that there is a two-year reproductive cycle (Benjaminsen & Christensen, 1979). Such a cycle could link up with the annual migration already described.

In species with gestation periods significantly longer than 12 months, it is difficult to postulate links with an annual migratory cycle. In the sperm whale the gestation period is about 15 months, in the white whale about 14.5 months, and in the long-finned pilot whale (*Globicephala melaena*) 15 to 16 months. The sperm whale and pilot whale are polygynous, and the white whale may be (Sergeant, 1962; Kleinenberg, Yablokov, Bel'kovich & Tarasevich, 1964; Fraker, 1980), though the evidence is equivocal. Best (1979) suggests that prolonged pregnancy, which effectively separates the breeding and calving seasons, is an adaptation to polygyny which, by reducing interference in the social structure of mixed schools during the calving period, reduces calf mortality.

Climate and physical environment

In polar waters the distribution of ice affects the seasonal distribution of cetaceans. In the arctic, the summer range of the white whale in some areas is ice covered from about October to June, and the timing of their spring migration is influenced by the break-up of the ice. In the autumn animals which remain too long in their summer range may be caught and trapped by advancing ice. Winter distribution is dependent upon the presence of open water.

The distribution and movements of the narwhal are similarly governed in part by ice distribution (Newman, 1978).

The distribution of some species of dolphins is related to sea surface temperature (Gaskin, 1968) but there is as yet no clear evidence that the migration patterns of odontocetes other than the white whale and narwhal are directly influenced by water temperature. Best (1979) notes that it is unlikely to be a limiting factor in the distribution of sperm whales and there is no clear evidence for this in the northern bottlenose whale (Benjaminsen & Christensen, 1979).

A special case of the effect of the environment is seen in the movements of the Ganges susu in the Ganges-Brahmaputra river system in India and Bangladesh. The dolphin is found in the main rivers and their tributaries during the rainy season (May to September) but most of the adults at least apparently retreat from the tributaries to the main rivers as the water levels decrease in the dry season from October to April (Kasuya & Haque, 1972).

Social organization of migration in odontocetes

Odontocete species appear to have a more complex social organization than mysticetes. As already noted, female sperm whales and small males migrate mainly within the boundaries of tropical and subtropical waters, while medium-sized and large males have a different migration pattern into temperate and polar waters. Best (1979) discusses the various theories proposed to account for this segregation of the larger males. He suggests that segregation, by increasing dispersion and reducing intraspecific competition for available resources, is advantageous to the species as a whole.

Information on the social organization of the northern bottlenose whale is limited (Benjaminsen & Christensen, 1979). It occurs most frequently in pairs or in small schools of three to six animals, but solitary whales are also common and larger groups are occasionally seen. Schools of two or three whales seem to be of the same sex and age (either immature or mature), while larger groups may include maturing whales of both sexes. Single animals may be immature animals of either sex, or old males. There is some evidence from catch statistics for the segregation of males from females in some areas, e.g., in Icelandic waters in May and June, and the catches here include a higher proportion of large males than elsewhere. The relation of this local segregation to the migration pattern of the species is unknown.

The white whale appears to have a complex social organization with segregation of different groups of animals both on the summering grounds and during migration. On the summering grounds where the herds congregate, pregnant and post-partum females with their calves may be segregated (Brodie, 1971). During migration the whales travel in groups of two or three to herds of hundreds of animals. The small groups often consist of an adult female with from one to three calves of different ages. Larger groups of 10–20 animals are often adult males only (Sergeant, 1962; Kleinenberg *et al.*, 1964). The relationship of these groupings to the migration pattern of the species is not yet clear.

Speed and distances in migration

Sperm whales have been observed from the air to swim at speeds of from two to five or six knots on the whaling grounds off Durban, South Africa (Gambell, 1968*b*). Clarke (1972) estimated that they can cover the 600 nautical miles from 40° S (the northern limit of distribution of antarctic cephalopods) to the Durban whaling grounds within eight days at a cruising speed of three to five knots. Kleinenberg *et al.* (1964) give a speed of one to five knots for migrating white whales. However, R. A. Davis & K. J. Finley (unpublished data) estimated the speed of migration of animals in Lancaster Sound, Canada, in September from consecutive observations of herds to average seven to ten knots for a 24-h period.

Direct evidence of the distances covered by sperm whales in their migrations comes from the capture of marked whales (Best, 1969). In the Southern Hemisphere a male marked in the antarctic in position 62°22′ S, 26°25′ E was shot off Durban in 31°41′ S, a minimum distance of about 1 800 nautical miles. A female marked in 5°58′ S, 7°37′ W covered at least 2 000 nautical miles before being killed in 33°00′ S, 16°32′ E. The greatest distance so far recorded is about 4 000 nautical miles covered by a male marked in the North Atlantic in 21° 33′ N, 17°55′ W and shot off South Africa four and a half years later in 33°20′ S, 16°52′ E.

Orientation and navigation

Access to cetaceans in the wild is difficult and there is as yet very little information on the mechanisms involved in their orientation and navigation during migration. Since the early 1960s the use of radiotelemetry in field studies of cetaceans has been developed and individuals of 13 species, including fin, humpback and gray whales, have now been radio tagged and successfully tracked over various distances for short periods of time (Leatherwood & Evans, 1979). Much new information on diving behaviour, range of movement, and the behaviour of animals in groups has resulted. Continuing studies, especially if combined with simultaneous sampling of the animals' environment, may provide information for the identification of the factors involved in orientation and navigation but the logistic problems of large-scale studies, particularly on the larger species, are formidable.

Baker (1978) and Norris (1968) have reviewed the wide variety of environmental clues and the possible mechanisms involved. Of these vision, echo-location, and sound production and communication may be mentioned. As already noted, gray whales migrate close inshore in shallow water along much but not all of their migration route along the Pacific coast of North America. Pike (1962) and Norris (1968) suggest that they recognize landmarks using a combination of visual recognition of coastal features seen in 'spying out' behaviour, and information on the bottom topography obtained during deep dives which occur especially at points where changes in course are made. In the case of the common dolphin (*Delphinus delphis*), Pilleri & Knuckey (1968) state that in the western Mediterranean Sea during July 1967 the schools of dolphins encountered were invariably swimming eastwards toward the rising sun in the morning, and westwards towards the setting sun in the late afternoon. They suggest the animals were orienting visually using the sun's position.

Many odontocetes produce sounds used in echo-location and this can provide the animals with information about their surroundings underwater. Evans (1974) has shown that the routes followed by radio-tagged common dolphins in coastal waters of southern California and Mexico are closely associated with prominent features of the sea-bed topography such as coastal escarpments and sea mounts. The distribution of *Delphinus* species off southern California is generally associated with areas in high topographic relief and the animals avoid plains and flat areas of the sea bed (Hui, 1979). The use of echo-location may therefore be an important feature of orientation in coastal waters.

The ability of both odontocetes and mysticetes to use sounds for communication suggests the possibility that social communication may be important in the orientation of schools of cetaceans travelling together. In mysticetes these sounds may be audible over considerable distances (tens to hundreds of miles) when propagated through the deep-sound channel in the ocean. Payne & Webb (1971) suggest that individual whales may in fact be in acoustic communication with each other over these distances.

References

Ash, C. E. (1953). Weights of antarctic humpback whales. *Norsk Hvalfangst-Tidende*, **42**, 387–91.

Baker, R. R. (1978). *The Evolutionary Ecology of Animal Migration*. London: Hodder & Stoughton.

Bannister, J. L. & Gambell, R. (1965). The succession and abundance of fin, sei and other whales off Durban. *Norsk Hvalfangst-Tidende*, **54**, 45–60.

Benjaminsen, T. (1972). On the biology of the bottlenose whale, *Hyperoodon ampullatus* (Forster). *Norwegian Journal of Zoology*, **20**, 233–41.

Benjaminsen, T. & Christensen, I. (1979). The natural history of the bottlenose whale, *Hyperoodon ampullatus* (Forster). In *Behavior of Marine Animals*, vol. **3**: *Cetaceans*, ed. H. E. Winn & B. L. Olla, pp. 143–64. New York: Plenum Press.

Best, P. B. (1969). The sperm whale (*Physeter catodon*) off the west coast of South Africa 4. Distribution and movements. *Division of Sea Fisheries Investigational Report*, No. 78, 1–12.

Best, P. B. (1979). Social organization in sperm whales, *Physeter macrocephalus*. In *Behavior of Marine Animals*, vol. **3**: *Cetaceans*, ed. H. E. Winn & B. L. Olla, pp. 227–89. New York: Plenum Press.

Brodie, P. F. (1971). A reconsideration of aspects of growth, reproduction and behavior of the white whale (*Delphinapterus leucas*), with reference to the Cumberland Sound, Baffin Island, population. *Journal of the Fisheries Research Board of Canada*, **28**, 1309–18.

Brown, S. G. (1954). Dispersal in blue and fin whales. *Discovery Reports*, **26**, 355–84.

Brown, S. G. (1977*a*). Whale marking: a short review. In *A Voyage of Discovery*, ed. M. Angel, pp. 569–81. Oxford: Pergamon Press.

Brown, S. G. (1977*b*). Some results of sei whale marking in the Southern Hemisphere. *Reports of the International Whaling Commission*, Special Issue **1**, 39–43.

Brown, S. G. (1978). Whale marking techniques. In *Animal Marking. Recognition Marking of Animals in Research*, ed. B. Stonehouse, pp. 71–80. London: MacMillan.

Budylenko, G. A. (1978). Distribution and migration of sei whales in the Southern Hemisphere. *Twenty-eighth Report of the International Whaling Commission*, pp. 373–7.

Bureau of International Whaling Statistics (1930–79). *International Whaling Statistics* Nos. 1–84. Oslo and Sandefjord.

Chittleborough, R. G. (1965). Dynamics of two populations of the humpback whale *Megaptera novaeangliae* (Borowski). *Australian Journal of Marine and Freshwater Research*, **16**, 33–128.

Christensen, I. (1979). Norwegian minke whales fishery in 1976 and 1977. *Twenty-ninth Report of the International Whaling Commission*, pp. 467–72.

Clarke, M. R. (1972). New technique for the study of sperm whale migration. *Nature*, **238**, 405–6.

Dawbin, W. H. (1966). The seasonal migratory cycle of humpback whales. In *Whales, Dolphins and Porpoises*, ed. K. S. Norris, pp. 145–70. Berkeley and Los Angeles: University of California Press.

Deacon, G. E. R. (1937). The hydrology of the Southern Ocean. *Discovery Reports*, **15**, 1–124.

Donnelly, R. G. (1969). Further observations on the southern right whale *Eubalaena australis*, in South African waters. *Journal of Reproduction and Fertility*, Supplement **6**, 347–52.

Doroshenko, N. V. (1979). Populations of minke whales in the Southern Hemisphere. *Twenty-ninth Report of the International Whaling Commission*, pp. 361–4.

Evans, W. E. (1974). Radio-telemetric studies of two species of small odontocete cetaceans. In *The Whale Problem. A Status Report*. ed. W. E. Schevill, pp. 385–94. Cambridge, Mass.: Harvard University Press.

Fraker, M. A. (1980). Status and harvest of the Mackenzie stock of white whales (*Delphinapterus leucas*). *Thirtieth Report of the International Whaling Commission*, pp. 451–8.

Gambell, R. (1968*a*). Seasonal cycles and reproduction in sei whales of the southern hemisphere. *Discovery Reports*, **35**, 31–134.

Gambell, R. (1968*b*). Aerial observations of sperm whale behaviour. *Norsk Hvalfangst-Tidende*, **57**, 126–38.

Gambell, R. (1972). Sperm whales off Durban. *Discovery Reports,* **35,** 199–358.

Gambell, R. (1973). Some effects of exploitation on reproduction in whales. *Journal of Reproduction and Fertility,* Supplement **19,** 533–53.

Gaskin, D. E. (1968). Distribution of Delphinidae (Cetacea) in relation to sea surface temperatures off eastern and southern New Zealand. *New Zealand Journal of Marine and Freshwater Research,* **2,** 527–34.

Gilmore, R. M. (1960). Census and migration of the California gray whale. *Norsk Hvalfangst-Tidende,* **49,** 409–31.

Hart, J. T. (1935). On the diatoms of the skin film of whales, and their possible bearing on problems of whale movements. *Discovery Reports,* **10,** 247–82.

Hjort, J., Lie, J. & Ruud, J. T. (1932). Norwegian pelagic whaling in the Antarctic. I Whaling grounds in 1929–1930 and 1930–1931. *Hvalrådets Skrifter,* No. 3, 1–37.

Holm, J. L. & Jonsgård, Å. (1959). Occurrence of the sperm whale in the Antarctic and the possible influence of the moon. *Norsk Hvalfangst-Tidende,* **48,** 161–82.

Hui, C. A. (1979). Undersea topography and distribution of dolphins of the genus *Delphinus* in the Southern California Bight. *Journal of Mammalogy,* **60,** 521–7.

International Whaling Commission (1979). Report of the Scientific Committee. *Twenty-ninth Report of the International Whaling Commission,* pp. 38–105.

Ivashin, M. V. & Rovnin, A. A. (1967). Some results of the Soviet whale marking in the waters of the North Pacific. *Norsk Hvalfangst-Tidende,* **56,** 123–35.

Kasuya, T. & Haque, A. K. M. A. (1972). Some informations on distribution and seasonal movement of the Ganges dolphin. *The Scientific Reports of the Whales Research Institute,* Tokyo, **24,** 109–15.

Katona, S., Baxter, B., Brazier, O., Kraus, S., Perkins, J. & Whitehead, H. (1979). Identification of humpback whales by fluke photographs. In *Behavior of Marine Animals,* vol. **3:** *Cetaceans,* ed. H. E. Winn & B. L. Olla, pp. 33–44. New York: Plenum Press.

Kawamura, A. (1970). Food of sei whale taken by Japanese whaling expeditions in the Antarctic season 1967/68. *The Scientific Reports of the Whales Research Institute,* Tokyo, **22,** 127–52.

Kawamura, A. (1974). Food and feeding ecology in the southern sei whale. *The Scientific Reports of the Whales Research Institute,* Tokyo, **26,** 25–144.

Kawamura, A. (1975). A consideration on an available source of energy and its cost for locomotion in fin whales with special reference to the seasonal migrations. *The Scientific Reports of the Whales Research Institute,* Tokyo, **27,** 61–79.

Kawamura, A. (1978). An interim consideration on a possible interspecific relation in Southern baleen whales from the viewpoint of their food habits. *Twenty-eighth Report of the International Whaling Commission,* pp. 411–20.

Kayser, C. (1961). The physiology of natural hibernation. *International Series of Monographs on Pure and Applied Biology. Modern Trends in Physiological Sciences* **8.** 325 pp. Oxford: Pergamon Press.

Kleinenberg, S. E., Yablokov, A. V., Bel'kovich, B. M. & Tarasevich,

M. N. (1964). *Beluga (Delphinapterus leucas) Investigation of the Species.* Moscow: Akademiya Nauk SSSR. Translation by Israel Program for Scientific Translations, Jerusalem, 1969.

Laws, R. M. (1959). The foetal growth rates of whales with special reference to the fin whale, *Balaenoptera physalus* Linn. *Discovery Reports,* **29,** 281–308.

Laws, R. M. (1961). Reproduction, growth and age of southern fin whales. *Discovery Reports,* **31,** 327–486.

Leatherwood, S. & Evans, W. E. (1979). Some recent uses and potentials of radiotelemetry in field studies of cetaceans. In *Behavior of Marine Animals,* vol. **3:** *Cetaceans,* ed. H. E. Winn & B. L. Olla, pp. 1–31. New York: Plenum Press.

Lockyer, C. (1972). The age at sexual maturity of the southern fin whale (*Balaenoptera physalus*) using annual layer counts in the ear plug. *Journal du Conseil,* **34,** 276–94.

Lockyer, C. (1974). Investigation of the ear plug of the southern sei whale, *Balaenoptera borealis,* as a valid means of determining age. *Journal du Conseil,* **36,** 71–81.

Lockyer, C. (1976). Growth and energy budgets of large baleen whales from the southern hemisphere. *Scientific Consultation on Marine Mammals,* Bergen. Paper ACMRR/MM/SC/41 (in press).

Lockyer, C. (1977). A preliminary study of variations in age at sexual maturity of the fin whale with year class, in six areas of the Southern Hemisphere. *Twenty-seventh Report of the International Whaling Commission,* pp. 141–7.

Lockyer, C. (1978*a*) The history and behaviour of a solitary wild, but sociable, bottlenose dolphin (*Tursiops truncatus*) on the west coast of England and Wales. *Journal of Natural History,* **12,** 513–28.

Lockyer, C. (1978*b*). A theoretical approach to the balance between growth and food consumption in fin and sei whales, with special reference to the female reproductive cycle. *Twenty-eighth Report of the International Whaling Commission,* pp. 243–9.

Lockyer, C. (1979*a*). Changes in a growth parameter associated with exploitation of southern fin and sei whales. *Twenty-ninth Report of the International Whaling Commission,* pp. 191–6.

Lockyer, C. (1979*b*). Review (in minke whales) of the weight/length relationship and the Antarctic catch biomass, and a discussion of the implications of imposing a body length limitation on the catch. *Twenty-ninth Report of the International Whaling Commission,* pp. 369–74.

Mackintosh, N. A. (1942). The southern stocks of whalebone whales. *Discovery Reports,* **22,** 197–300.

Mackintosh, N. A. (1966). The distribution of southern blue and fin whales. In *Whales, Dolphins and Porpoises,* ed. K. S. Norris, pp. 125–44. Berkeley and Los Angeles: University of California Press.

Mackintosh, N. A. (1972). Life cycle of Antarctic krill in relation to ice and water conditions. *Discovery Reports,* **36,** 1–94.

Mackintosh, N. A. (1973). Distribution of post-larval krill in the Antarctic. *Discovery Reports,* **36,** 95–156.

Marr, J. W. S. (1956). *Euphausia superba* and the Antarctic surface currents. *Norsk Hvalfangst-Tidende,* **43,** 127–34.

Marr, J. (1962). The natural history and geography of the Antarctic krill (*Euphausia superba* Dana). *Discovery Reports,* **32,** 33–464.

Masaki, Y. (1979). Yearly change of the biological parameters for the Antarctic minke whale. *Twenty-ninth Report of the International Whaling Commission,* pp. 375–98.

Mercer, M. C. (1975). Modified Leslie-DeLury population models of the long-finned pilot whale (*Globicephala melaena*) and annual production of the short-finned squid (*Illex illecebrosus*) based upon their interaction at Newfoundland. *Journal of the Fisheries Research Board of Canada,* **32,** 1145–54.

Mitchell, E. & Kozicki, V. M. (1975). Autumn stranding of a northern bottlenose whale (*Hyperoodon ampullatus*) in the Bay of Fundy, Nova Scotia. *Journal of the Fisheries Research Board of Canada,* **32,** 1019–40.

Nemoto, T. (1959). Food of baleen whales with reference to whale movements. *The Scientific Reports of the Whales Research Institute,* Tokyo, **14,** 149–290.

Nemoto, T. (1962). Food of baleen whales collected in recent Japanese Antarctic whaling expeditions. *The Scientific Reports of the Whales Research Institute,* Tokyo, **16,** 89–103.

Nemoto, T. (1963). Some aspects of the distribution of *Calanus cristatus* and *C. plumchrus* in the Bering and its neighbouring waters, with reference to the feeding of baleen whales. *The Scientific Reports of the Whales Research Institute,* Tokyo, **17,** 157–70.

Newman, M. A. (1978). Narwhal. In *Marine Mammals of Eastern North Pacific and Arctic Waters,* ed. D. Haley, pp. 139–144. Seattle: Pacific Search Press.

Nishiwaki, M. & Sasao, A. (1977). Human activities disturbing natural migration routes of whales. *The Scientific Reports of the Whales Research Institute,* Tokyo, **29,** 113–20.

Norris, K. S. (1968). Some observations on the migration and orientation of marine mammals. In *Animal Orientation and Navigation,* ed. R. M. Storm, pp. 101–25. Corvallis: Oregon State University Press.

Ohsumi, S. & Masaki, Y. (1975). Biological parameters of the Antarctic minke whale at the Virginal population level. *Journal of the Fisheries Research Board of Canada,* **32,** 995–1004.

Omura, H. (1950). Diatom infection on blue and fin whales in the Antarctic whaling area V (the Ross Sea Area). *The Scientific Reports of the Whales Research Institute,* Tokyo, **4,** 14–26.

Omura, H. (1973). A review of pelagic whaling operations in the Antarctic based on the effort and catch data in 10° squares of latitude and longitude *The Scientific Reports of the Whales Research Institute,* Tokyo, **25,** 105–203.

Omura, H. & Ohsumi, S. (1964). A review of Japanese whale marking in the North Pacific to the end of 1962, with some information on marking in the Antarctic. *Norsk Hvalfangst-Tidende,* **53,** 90–112.

Payne, R. (1976). At home with right whales. National Geographic, **149,** 322–39.

Payne, R. & Webb, D. (1971). Orientation by means of long range acoustic signalling in baleen whales. *Annals of the New York Academy of Sciences,* **188,** 110–42.

Pike, G. C. (1962). Migration and feeding of the gray whale (*Eschrichtius gibbosus*). *Journal of the Fisheries Research Board of Canada,* **19,** 815–38.

Pilleri, G. & Knuckey, J. (1968). The distribution, navigation and orientation by the sun of *Delphinus delphis* L. in the western Mediterranean. *Experientia*, **24**, 394–6.

Rice, D. W. (1965). Offshore southward migration of gray whales off southern California. *Journal of Mammalogy*, **46**, 504–5.

Rice, D. W. & Wolman, A. A. (1971). *The Life History and Ecology of the Gray Whale (Eschrichtius robustus)*. The American Society of Mammalogists, Special Publication No. 3, 142 pp.

Sergeant, D. E. (1962). The biology and hunting of beluga or white whales in the Canadian Arctic. *Fisheries Research Board of Canada, Arctic Unit*, Circular No. 8, 13 pp.

Sergeant, D. E. & Fisher, H. D. (1957). The smaller Cetacea of eastern Canadian waters. *Journal of the Fisheries Research Board of Canada*, **14**, 83–115.

Slijper, E. J. (1962). *Whales*. London: Hutchinson.

Taylor, R. J. F. (1957). An unusual record of three species of whale being restricted to pools in Antarctic sea-ice. *Proceedings of the Zoological Society of London*, **129**, 325–31.

Townsend, C. H. (1935). The distribution of certain whales as shown by logbook records of American whaleships. *Zoologica*, **19**(1), 1–50.

Würsig, B. & Würsig, M. (1977). The photographic determination of group size, composition and stability of coastal porpoises (*Tursiops truncatus*). *Science*, **198**, 755–6.

F.R.HARDEN JONES

Fish migration: strategy and tactics

Introduction

A man emigrates when he quits one country, and immigrates when he enters another to settle: these are one-way trips and the traveller needs but a single ticket. The word migration is often used in the sense of moving from one place to another. But for animals which live for a year or more migration has a special meaning: there is a coming and going with the seasons – even if not always on an annual basis – and the traveller needs a return ticket. This is the meaning of the word that I have in mind when referring to the migrations of fish.

There are about 8 000 species of fish which live in freshwater and a further 12 000 which live in the sea; and there are about 120 diadromous species which move regularly between the two environments (Cohen, 1970). Fish are not distributed evenly throughout the world. About 100 species are found within the Antarctic convergence (Marshall, 1964) which suggests that there may not be more than 200 or 300 species in polar latitudes. There are many more species in temperate waters but the majority – probably at least 70% of the total – are found within the tropics where over 500 species may be found on a single reef.

Many fish are restricted in that they are limited to one geographical area within which they make wintering, spawning, and feeding migrations. These movements are often relatively short. But some species, perhaps no more than a few hundred, include stocks* which migrate between widely separated areas and here the distances involved may be substantial. The distinction is between distances of 20 to 30 km and those of several hundreds or even a few thousand kilometres.

It is the relatively few species with stocks that make extensive migrations that have hitherto attracted most attention. There are two reasons for this. Firstly the Great Fisheries depend on them; and secondly the length, regularity, and precision of the seasonal movements excites the imagination. In what follows I will touch upon the biological background to the first point and explore two of the behavioural problems of the second: the first point is concerned with strategy, and the second with tactics.

---

*The unit stock has been defined as 'a relatively homogeneous and self-contained population whose losses by emigration and accessions by immigration, if any, are negligible in relation to the rates of growth and mortality' (Anon, 1960, p. 8).

The migration patterns in relation to strategy

Migratory behaviour is one of several features in the life histories of fish directed towards reproductive success. Here the Grand Strategy appears to be that of producing a sufficient number of viable offspring to maintain the population up to the limit – in numbers or in weight – that can be fed. Nikolsky (1963, p. 235) regards migration as 'an adaptation towards increasing the abundance of a species' and this is a view with which I have some sympathy. But Cushing (1969, p. 210) puts the question as to 'whether fish are abundant because they migrate or *vice-versa*.' I do not know of any simple argument to resolve this circularity and when others have tried to do so, the results have not been very constructive (Mailyan, 1970; Svetovidov, 1970). Perhaps it is sufficient to make two points. Firstly, the food resources of different areas would not be available to stocks unless they moved between them; and secondly, those groups which are the more important commercially – cods, herrings, tunas, and salmons – are strongly migratory. According to the FAO Statistical Yearbook (FAO, 1978), in 1977 the world catch of marine fish amounted to 54 million tons. While 500–600 species were included in the total, over half the catch was contributed by 25 species and among these only one – the sand eel, *Ammodytes marinus* – could be fairly described as being non-migratory. The sand eel ranks tenth among the Top Species listed in Table 1.

Table 1. *World catch of marine fish in 1977*

Species	Metric tons $\times 10^{-3}$	Cumulative % total
1 Alaskan pollock	4 296	8
2 Capelin	4 009	15
3 Atlantic cod	2 298	20
4 Japanese mackerel	2 030	23
5 Japanese sardine	1 470	26
6 Chilean pilchard	1 468	29
7 Atlantic herring	994	31
8 Jack mackerel	840	32
9 Peruvian anchovy	807	34
10 Sand eels	804	35
Cumulative % totals		
First 24 species		50
First 53 species		62

Note: The top ten species and their contribution to the total marine catch of 54 million tons. The cumulative % totals for the first 24 and 53 species are also given. (Data from FAO, 1978.)

One consequence of migration is that a species, or a particular stock, may enjoy a greater food supply and individuals may therefore grow faster and have a higher chance of survival, particularly among the larval and younger stages. A faster growth rate would allow more juveniles to escape predation because they would spend less time at risk in a vulnerable size range; the better-fed survivors might so increase their length-at-age as to mature a year or more earlier than normal; and those mature individuals which enjoy an abundant food supply could produce more eggs up to a predetermined limit for the species and then go on to increase the reserves of a proportion or all of the eggs released.

Increased survival among the larval and younger stages, earlier maturity, and an increase in relative fecundity – the number of viable eggs per unit of body weight – will all contribute to an increase in the population. And migration allows a particular stock – if the population is so divided – to exploit, on a seasonal basis, the resources of different areas.

If numbers are so dependent on the availability of food, the migration patterns and the life histories of the fish should be geared to, or closely related to, the regional production cycles and regional hydrographic circulation. There is a lot of evidence to show that this is indeed so (Cushing, 1975).

In temperate waters the movements of migratory fish can be represented by a simple triangular pattern in which the movements of the young and adults are linked together. In general terms these movements are related to the water currents and the young stages drift passively from the spawning area to the nursery area where they grow up. In the traditional theory of fish migration the pre-spawning movements of the adult fish are described as an active contranatant migration *against* a prevailing current; and the return of the spent fish as a passive denatant movement *with* the prevailing current. Problems of behaviour and directionality were thus avoided, if not solved, in a broad but not entirely satisfactory generalization. For pelagic species, at least, it is not clear how the prespawning fish detect and orientate to the prevailing current when in midwater and so both out of sight and out of touch of the bottom.

As the adult fish feed in one area, winter in a second, and spawn in a third, their migratory movements can also be represented by a triangular pattern: the sequence of seasonal movements around the migratory circuit is from feeding area-to-spawning area-to-winter area for autumn spawners, and from feeding area-to-winter area-to-spawning area for spring spawners. Some examples follow.

Atlanto–Scandian herring

This stock of herring (*Clupea harengus*) is contained within the basin of the Norwegian Sea where there is a partly cyclonic circulation to which the warm North Atlantic Drift and the cold East Icelandic current (Figs. 1, 2) contribute the northwesterly and southeasterly legs respectively. Herring lay demersal eggs and the Atlanto–Scandian stock spawns in spring off the southwest coast of Norway. After hatching the pelagic larvae drift with the north-going current and the young fish spend their first two years in coastal waters. The older and mature fish are oceanic. After spawning they move north and west towards their summer feeding area whose western boundary lies along the line of the productive front between warm Atlantic and cold Arctic waters from Jan Mayen to Spitsbergen. The position of the front appears to be related to that of the sub-

Fig. 1. North Atlantic place names and depth contours relevant to the migrations of the Atlanto–Scandian herring and Arcto–Norwegian cod. The main herring-spawning areas off the southwest coast of Norway, and that of the cod within the West Fjord, are indicated.

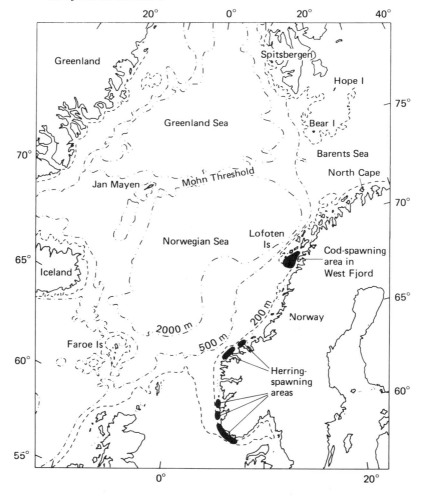

marine ridge – the Mohn Threshold – which separates the basins of the Norwegian and Greenland Seas in this area. In autumn the herring move to the southwest along the western boundary of the Norwegian Sea and subsequently to the southeast in the East Icelandic current. In winter the herring concentrate in relatively cold water (3–4°C) and at depths of 300 to 400 m, to the north of the Faroes. In March the ripe herring reach the coastal spawning grounds and so complete an annual migration of over 3000 km.

The herring move anticlockwise round the circuit and while there is no detailed evidence to show how the direction of swimming compares with that of

Fig. 2. The main surface currents in the North Atlantic relevant to the migrations of the Atlanto–Scandian herring and the Arcto–Norwegian cod.

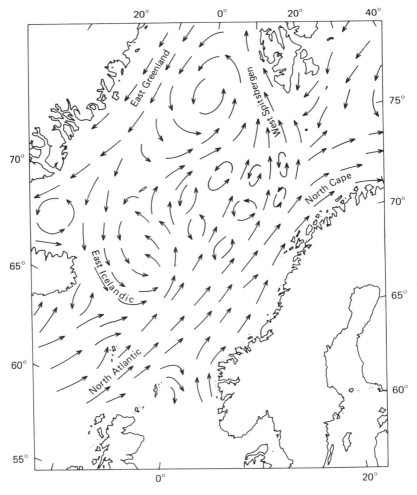

the water, the overall migration is more consistent with a movement down-stream, and so with the current, rather than against it.

The triangular pattern of migration is readily recognizable among other species in Arctic and temperate waters, but may be distorted by compression and elon-gation in stocks whose distribution is confined to the continental shelf which narrows in lower latitudes. Two examples, one from the Atlantic, and the other from the Pacific, are given below.

Arcto–Norwegian cod

The cod, *Gadus morhua,* has a wide distribution in the Atlantic. There are separate stocks at Georges Bank, Newfoundland, Labrador, West Greenland, East Greenland and Iceland, Faroe, Faroe Bank, in the Irish Sea, the North Sea (two or more stocks), the Baltic, and the Barents Sea. The cod in the Barents Sea – the Arcto–Norwegian stock – have as their summer feeding area the exten-sive shelf bounded by the 200-m depth contour within which important place names are Bear Island, Spitsbergen, Hope Island, and North Cape (see Fig. 1). In the autumn the distribution of the cod contracts and in winter dense concen-trations build up on the bottom near Bear Island in association with the front between warm Atlantic and cold Arctic water. Later in the year the mature fish, accompanied for some of the way by the larger immatures, leave the Bear Island Bank for the Norwegian coast. Only the mature fish go on to their spawning grounds which extend from North Cape to Romsdal, the most important area being in the arm of the West Fjord, within the Lofoten Islands. After spawning the spent cod return to their northerly feeding area, while the pelagic eggs and larvae are carried to the north in the coastal current.

Alaskan pollock

The Alaskan pollock (or pollack or walleye), *Theragra chalcogramma,* is a whiting whose distribution in the North Pacific parallels that of the cod in the North Atlantic. There are stocks of pollock in the Sea of Japan, Sea of Okhotsk, Bering Sea, Alaskan Coastal waters, off Vancouver Island and in American coastal waters as far south as California. The walleye pollock, like the Atlantic cod, is a shelf fish. In the eastern Bering Sea the distribution of the pollock is confined between the western edge of the shelf (indicated in Fig. 3 by the 500-m depth contour) which runs from Unimak Island to Cape Navarin, and to the east by water of less than 2°C in winter and the 50-m depth contour in summer.

To the north of Unimak Island, pollock form dense concentrations in winter along the lines of the 150–280-m depth contours where bottom water tempera-tures range from 2.5 to 4.5°C (Serobaba, 1970). From March to June spawning occurs to the north of Unimak Pass at depths of 100 to 300 m (Serobaba, 1968).

Fig. 3. Place names in the Bering Sea relevant to the migrations of the Alaskan pollock. The pollock spawning area to the northwest of Unimak Island is indicated.

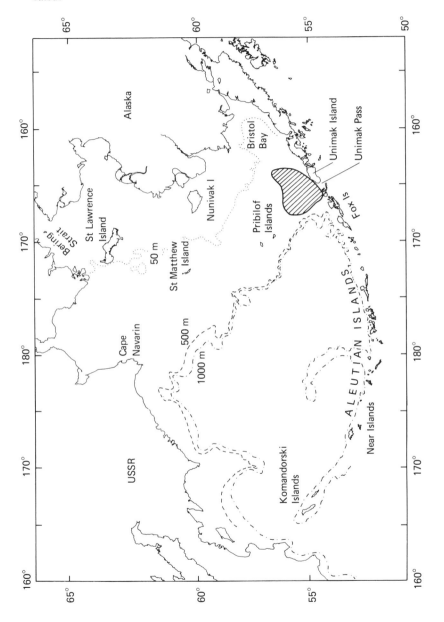

Peak spawning occurs in May and the pelagic eggs, larvae, and young fish are carried towards and contained within the shallow shelf waters by local current systems (Fig. 4). During the summer the pollock migrate to the west and north and then spread over the shelf past the Priblof Islands toward St Matthew Island and beyond. At the end of the autumn the distribution of the pollock contracts and there is a return to the deeper waters on the edge of the shelf with the onset of winter.

The Arcto–Norwegian cod and the pollock of the Northeastern Bering Sea show remarkable similarities in their annual migration circuits. Both stocks winter towards the edge of a shelf in deep and relatively warm water; the spring spawning areas are well to the south of the summer feeding areas which are themselves spread across an extensive shelf; and the migrations and distribution of both stocks would appear to be related to movements of two warm-water currents, the North Atlantic–West Spitsbergen current and the Bering Slope current (which flows to the northwest from Unimak to Cape Navarin along the line of the shelf) respectively. Indeed, the similarities between the hydrography of the two areas, which extends to local within-stream eddies and deep counter-currents, has been noted by physical oceanographers (Kinder, Coachman & Galt, 1975).

Fig. 4. Main surface currents in the Bering Sea compiled from various charts in Hughes, Coachman & Aagaard (1974) and Favorite, Dodimead & Nasu (1976, Fig. 42). The migrations of older Alaskan pollock are related to the Bering Sea Slope current and the distribution of the larvae to the West Alaska current. The spawning area is indicated on Fig. 3.

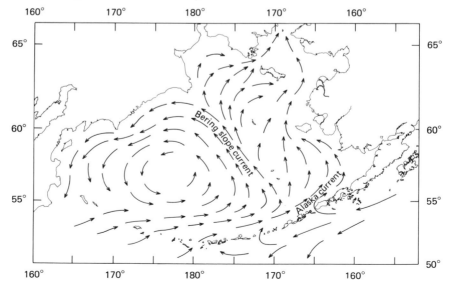

Tactics and behaviour

If asked to select keywords or phrases to characterize the salient features of fish migration I would choose hydrographic containment, regularity of movement, and homing. Furthermore, I would argue that these features are related to the seasonal production cycles in temperate and arctic water, and in lower latitudes to the essentially similar production cycles in areas of divergence and in particular the upwelling areas associated with the eastern boundary currents – the California, Peru, Canary, and Benguela. Of these three salient features, it is the study of homing that has been most neglected in marine fish and here I am moving from the biological towards the behavioural problems of migration.

Homing

The Olfactory Hypothesis and salmonids Homing – following Gerking's (1959) definition as 'the return to a place formerly occupied instead of going to other equally probable places' – has been most clearly demonstrated in salmons. One of the most convincing set of data is that provided by R. E. Foerster who worked with the sockeye salmon, *Oncorhynchus nerka,* in Cultus Lake, on the Fraser River, British Columbia. Foerster (1937) fin clipped many thousands of smolts leaving the lake. Marked fish subsequently recovered as mature salmon in the Fraser River – albeit less than 10% of the original downstream migrants – returned unerringly to their native lake. While later work with other stocks of salmonids have shown that straying does occur, the rates are low, certainly less than 10% among those fish that return to freshwater to spawn, and sometimes lower than 5%.

The sensory basis for homing in salmonids is at least partially understood and A. D. Hasler is associated with the Olfactory Hypothesis which has as its cornerstone the suggestion 'that juvenile salmon become imprinted to the unique chemical odour of their natal stream during the smolt stage and subsequently use this cue to locate their stream during the spawning migration' (Hasler & Scholz, 1980, p. 179). Hasler & Scholz have reviewed, in detail, the experimental evidence to support the Olfactory Hypothesis, and while there still seems to be some doubt as to the role that olfaction plays in homing behaviour (Hara & Brown, 1979), the importance of imprinting does not appear to be in dispute.

One interesting point in the Olfactory Hypothesis is the suggestion that imprinting can still take place relatively late in the life history during the smolt stage in the second, third, or even fourth year of life. I have often wondered if a similar mechanism could be involved in the return of marine fish to their spawning area or ground: for it seems likely that some local landmark must be involved if only because fish spawn at one locality rather than another.

Marine species and the Groundwater Seepage Hypothesis To stimulate discussion I want to mention the Groundwater Seepage Hypothesis which I have developed elsewhere (Harden Jones, 1980). The essential feature of this new hypothesis is that the assembly areas, spawning areas, and spawning grounds of marine fish which spawn in coastal waters could be identified by reference to chemicals entering the sea by groundwater seepage. The hypothesis has some conceptual merit as it could link the anadromous species of some orders with their fully marine counterparts and would extend Hasler's hypothesis to the sea. Specifically I am suggesting that plaice, cod, herring, pilchard, mackerel, and other marine species which spawn on or above the continental shelf are homing to areas where there is a seepage of groundwater, perhaps on only a small scale, from a cluster or line of submarine springs with individual flow rates of a few tens of litres per second.

One of the geological conditions under which groundwater seepage could occur under the sea is not exceptional (see Fig. 5) and the idea itself is certainly not new, having been suggested by Robert Stephenson in 1840 and later endorsed by Boyd Dawkins (1898).

Evidence for submarine springs is documented for both the western (Emery & Hulsemann, 1963, p. 37) and eastern (Manheim, 1967; Bokuniewicz, 1980) coastal waters of North America. Springs are present in the Persian Gulf off

Fig. 5. The discharge of groundwater into the sea could take place under the conditions shown in the diagram. 1, superficial permeable deposits; 2, impervious layer of clay; 3, chalk aquifer saturated to indicated level. A, medium discharge through a pit; B, minor discharge through a crack or fissure; C, major discharge through an outcrop. An artesian head of about three metres would probably be sufficient to allow fresh groundwater to enter the sea at a depth of 100 m (density of seawater 1.026 g cm^{-3}).

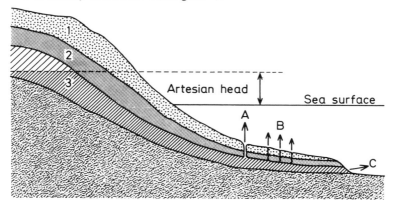

Bahrain (Tibbetts, 1971). Huxley (1894, p. 67) notes the presence of submarine springs in the Mediterranean where they occur off Cap-Martin (Idrac, 1933), in the Adriatic (Buljan & Zore-Amanda, 1976), and elsewhere (Kohout, 1966). And those who know their Coleridge may recall the lines

> Where Alph, the sacred river, ran
> Through caverns measureless to man
> Down to a sunless sea.

Returning from Ionian to United Kingdom waters, there are foreshore outlets along the coast from Flamborough Head to Rottingdean (Davies, 1973; Brereton & Downing, 1975), and Boyd Dawkins (1898, p. 264) considered that 'outlets on the foreshore are probably insignificant compared with those which may reasonably be expected to exist below the level of the sea where they cannot be examined'.

While chalk could be a source of groundwater seepage under the sea in the southern and eastern region of the British Isles, Karstic drainage from limestone is a likely source in other areas, for example in Galway and Morecambe Bays, off the coast of South Wales, and perhaps in the western English Channel. While a systematic search for low salinity bottom water has not yet been made in any British coastal area, freshwater discharges certainly occur. Storer (1815) records the presence of freshwater below Bridlington Harbour which was thought to reach the sea through fissures in the rock on the eastern side of Smithic Sand; there are the 'Hessle Whelps' in the River Humber (Versey, 1946); Mr G. R. Forster (personal communication) found a freshwater spring while diving in the Tor Bay area; and in 1950 Dr L. H. N. Cooper (personal communication) collected anomalous low salinity bottom samples on the Nymphe Bank (southeast Ireland) which were thought to indicate groundwater seepage.

The Groundwater Seepage Hypothesis would require that both nursery and spawning areas received freshwater from the same aquifer or drainage system. If imprinting in marine species was delayed until the young fish were in their inshore nursery area they might be sensitized to some chemical present in groundwater entering the sea through foreshore outlets. Young fish usually leave shallow water at the end of their first summer, and it is therefore possible that they might not come across the chemical substance to which they had been originally imprinted until, as mature fish, they arrived within the immediate vicinity of the spawning area. Nursery and spawning area would thus be linked not only geographically in terms of residual currents and larval drift, but also in terms of geological structure and groundwater.

Some of you might consider that I have gone too far ahead of the facts and you may be sceptical of the whole idea: I too, have my doubts and would plead, in mitigation, Huxley's Defence*. Perhaps one needs to be graced with a moment of serendipity and here I have been fortunate as friends have drawn my attention to several instances where marine species are unexpectedly associated with freshwater. Thus Richard Carew, writing in 1602 of the pilchard in Cornish waters, noted that 'they come to take their kind of the fresh (as the rest) between harvest and Allhallowtide . . .' (Carew, 1811, p. 100); Sars (1879, p. 571) in a report on the Lofoten cod fishery, wrote of 'the idea very generally prevailing among fishermen that the codfish seek certain places on the bottom where there are said to be springs of fresh water which they drink in order to bring the roe to maturity'; Huntsman (1934, p. 85) reported that in Passamaquoddy 'there is a belief held by some fisherman that "herring go to freshwater to get a drink" ', which recalls Smitt's (1895, p. 966) comment that the herring 'chooses its spawning-places in water shallower than its ordinary home, and during its youth at least, it also seeks for this purpose water of less salinity'; and Forster (1974) has described the marine coelocanth *Latimeria* as inhabiting caves into which there is a seepage of freshwater. My colleague, Dr T. Wyatt, has drawn my attention to Le Danois' (1934, Fig. 34) chart of the old submerged courses of the Severn and the Seine which reach the edge of the continental shelf in areas where mackerel and pilchards now spawn. Le Danois (1934, pp. 438–9) notes 'c'est justement de long de ces thalwegs géologiques que se produisent les déversements des eaux continentales; ces eaux lourdes de basse température et de faible salure, suivent le cours ancien de la Seine et de la Severn'. This suggests the possibility of a relation between old river courses, groundwater and present spawning sites.

Whilst I would agree that this is perhaps going too far even for Huxley, it serves to highlight the fact that we know very little about why fish select one area, or ground, for spawning rather than another. The Atlanto–Scandian spring spawning herring, to which I have referred earlier, provide an example. In the 1930s these herring spawned off the west coast of Norway in two main areas: to the north off More, between Stadt and Kristiansund; and to the south off Rogalund, between Bergen and Lindesnes (see Fig. 1). Runnström (1941a,b) investigated the spawning of the Norwegian spring herring from 1931 to 1937. He showed that the age composition of the shoals changed during the spawning season and that the younger herring, the first-time recruit spawners, were the last to arrive (Runnström, 1941a, Figs. 10a and 10b). For several years he charted the distribution of the demersal herring eggs with a network of Petersen grab

---

* '. . . anyone who is practically acquainted with scientific work is aware that those who refuse to go beyond fact, rarely get as far as fact; and anyone who has studied the history of science knows that almost every great step therein has been made by the "anticipation of Nature", that is by the invention of hypothesis, which, though verifiable, often had very little foundation to start with; and, not infrequently, in spite of a long career of usefulness, turned out to be wholly erroneous in the long run' (Huxley, 1893, p. 62).

samples and in 1937 supported these observations with the first systematic echo survey. The southern spawning area was studied in detail. Within this area there are several banks with rocky or sandy bottoms separated from each other by deeper channels with soft bottoms. On the basis of the grab samples Runnström (1941b) identified 14 banks as being favourable spawning grounds and the positions of these banks is shown in Fig. 6. The spawning date of the eggs was estimated from their developmental stage and in this way Runnström was able to show the sequence of spawning over the 14 grounds. Spawning started to the southwest of Karmoy on grounds 8, 7 and 9, and later spread to the south (grounds 10 to 14) and to the north (grounds 6 to 1). This shift in spawning ground – here summarized in Table 2 – was consistent with the results of the echo surveys which suggested an inshore movement of herring shoals from the west which spread north and south when they reached the coastal banks. The most northerly ground – number 1 off Bommelen – is mainly visited by the recruit spawners (Runnström, 1941a, p. 26; 1941b, p. 35–8). This is curious as it implies that they spawn on another ground the following year. Does the presence of herring eggs on particular banks in some way deter further spawning –

Table 2. *Mean date of first spawning by Atlanto–Scandian herring in the Utsira–Karmoy–Egersund area*

Spawning ground	Mean date of first spawning
1	4 Mar
2	8 Mar
3	7 Mar
4	2 Mar
5	24 Feb
6	22 Feb
7	16 Feb
8	7 Feb
9	9 Feb
10	16 Feb
11	27 Feb
12	17 Feb
13	16 Feb
14	21 Feb

Note: Data from Runnström (1941 b, Fig. 9). The location of the spawning grounds 1 to 14 is shown in Fig. 6.

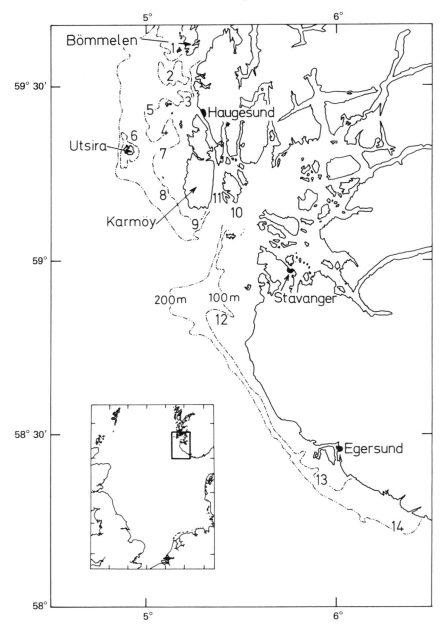

Fig. 6. Herring-spawning grounds off the southwest coast of Norway (Runnstrom, 1941*b*). The 100- and 200-m depth contours are indicated.

perhaps through the agency of a pheromone – so that the recruits, the last to arrive, end up at the extreme boundaries of the area? One might then expect that the most southerly grounds would also be visited mainly by recruits; but this does not appear to be so. And in 1936, although there was little or no spawning on grounds 2 to 10 – the herring having gone elsewhere owing to bad weather – the recruit spawners still appeared at Bommelen at the end of the season. We do not know why the fish chose one spawning ground rather than another, and research on this problem is needed.

Orientation and migration

Arnold (this volume) describes some of our recent work with plaice. In the Southern Bight of the North Sea these flatfish appear to use the tidal streams for transport when moving between their feeding and spawning areas, and they come off the bottom to join the appropriate tide. We have always wanted to go a step further and determine the direction in which a plaice was heading when in midwater. With a density of 1.063 g cm^{-3} (Lowndes, 1941) a plaice would have to swim through the water in order to maintain its height above the bottom, but it would not be expected to maintain a steady course in the absence of visual or tactile clues. Nevertheless, when simple vector analysis was applied to those midwater tracks for which current meter data were available, the results showed that some fish maintained a surprisingly consistent course for periods up to two hours (Greer Walker, Harden Jones & Arnold, 1978, Figs. 5, 11 and 17).

Our colleagues at the Fisheries Laboratory, Lowestoft, have invented an acoustic transponding compass tag (R. B. Mitson, T. R. Storeton West and N. D. Pearson, in preparation) so that we can now obtain a virtually continuous record of the position, depth, and heading of a fish in real time. The new tag, which is somewhat larger than the conventional acoustic transponder, is attached to the plaice by two Petersen discs and so aligned along the body axis on the upper right-hand side. The compass reference is provided by a magnet attached to a slotted disc and mounted on a jewelled pivot. Changes in direction are sensed optically using a circular array of 8 infrared emittors and 8 detectors to determine the bearing of the slot – and so the heading of the fish – which can be assigned to one or other of 8 nominally 45° sectors (Fig. 7). When the tag is interrogated at 300 kHz, it responds with a reference pulse which can be used to determine the position and the depth of the fish in the usual way (Greer Walker et al., 1978). The compass heading is indicated by a second pulse, the delay between the two signals (ranging from 24 to 66 ms in seven 6-ms steps) identifying the relevant sector. The two pulses are shown on the B-scan sonar display and, from selected channels of the receiving beam, on an Alden paper recorder (see Fig. 8). The heading of the fish can also be read directly from a digital decoder.

The track of a plaice fitted with a compass tag and released in the open sea 24 km off Orfordness is shown in Fig. 9. The fish was released at 1730 h GMT on 3 June and abandoned at 2400 h on 5 June. It left the bottom during the night and came into midwater for 5 periods as indicated in Fig. 10 and Table 3. Neither moon nor stars were visible on these nights. When in midwater the plaice was

Fig. 7. The characteristics of the compass tag fitted to the plaice released on 3 June 1979.

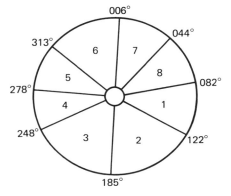

Fig. 8. Alden recorder chart showing the range and compass heading of the fish from 2300 to 2320 h on 3 June 1979 during midwater period 1. The compass sector headings are indicated. The marks extending in range and labelled V indicate a change from horizontal to vertical mode to check the depth at which the fish was swimming.

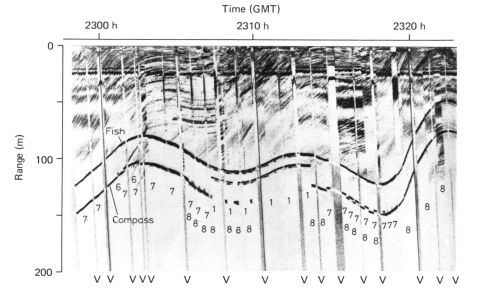

Table 3. *Data relating to the track of a plaice fitted with an acoustic transponding compass tag released off Orfordness on 3 June 1979*

| Period in midwater | Fish | | | | Tide | |
| | Time in hours, GMT | | Movement over ground | | | |
	off bottom	on bottom	direction °T	speed cm s⁻¹	Direction °T	Speed cm s⁻¹
1	2207	2352	012°	43	N 009°	39
2	0200	0242	175°	95	S 198°	54
3	2046	2338	040°	60	N 015°	62
4	0040	0302	166°	38	S 190°	65
5	2056	2400	038°	66	N –	–

Note: The track of the fish is shown in Fig. 10. The tidal velocity was estimated from moored current meters (see Fig. 9) which were recovered between midwater periods 4 and 5.

Fig. 9. The track of a plaice fitted with an acoustic transponding compass tag and released at 1730 h GMT on 3 June 1979. The 10-, 20-, and 30-m depth contours and the positions of two moored current-meter stations are indicated.

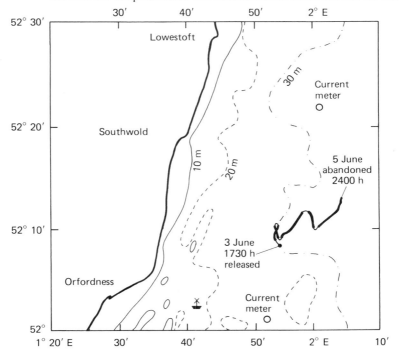

usually from 5 to 20 m above the bottom (total depth 25–30 m) to which it returned several times. These excursions to the bottom rarely exceeded two to three minutes.

Fig. 11 shows the headings, by sectors, for each of the five midwater periods: they were not randomly distributed. In period 1 the fish was usually heading in sectors 6 and 7 (313° to 044°) and in periods 2 to 5 in sectors 1 and 2 (082° to 185°). The courses made good over the ground during periods 1 and 3 are markedly different (012° and 040°) although in both instances the direction of the tide was virtually the same (009° and 015°). The difference between the two ground courses would appear to be related to the sector headings maintained by the fish in midwater: in period 1 the fish was heading downtide and in period 3 the fish was heading across the tide. The ground course of 040° appears to be the resultant of the fish's velocity through the water and that of the tide. The results suggest that a fish in midwater will actively swim and maintain a fairly consistent heading in one direction while being carried by the tide in another.

The distribution patterns shown in Fig. 11 are still evident when each midwater period is considered in parts separated by bottom excursions. Two examples are given in Figs. 12 and 13 which show that the patterns do not

Fig. 10. Details of the track chart of a plaice fitted with an acoustic transponding compass tag and released at 1730 h GMT on 3 June 1979. The five midwater periods are indicated by bold lines. The plaice was abandoned at 2400 h on 5 June.

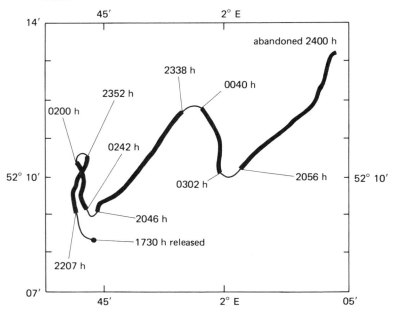

vary with average height above the bottom or with the time spent in midwater.

We have looked at the data in greater detail to get some measure of the length of time for which a plaice could maintain its heading when in midwater. With the rather coarse compass reference available the best that could be done was to measure, on the Alden charts, the time that the fish remained in one sector before being recorded in another. The results are summarized in Table 4. There were many instances when the fish was recorded as remaining in one sector for less than a minute and at the other end of the scale, there was a single instance when the fish remained heading in one 45° sector (which suggests that the heading was consistent to ± 22.5°) for just over 14 min. When the data were examined on the less stringent basis of consistently keeping in two adjacent sectors, there was one instance when the plaice appeared to maintain a heading to within ± 45° for 58 min (Table 5).

These results were unexpected. Furthermore, the data may underestimate the ability of a plaice to maintain a steady course as there were instances where the compass signals were regularly alternating between two adjacent sectors which suggested that the fish was keeping to a narrow range of bearings towards the edge of either sector.

If further studies confirm these first results, we are going to have to deal with two new problems: firstly, what is the reference used to select the course when

Fig. 11. The heading of a midwater plaice as indicated by the compass tag for each of the five midwater periods shown in Fig. 10 and summarized in Table 3. The time recorded in each sector is expressed as a proportion of the total time for which the fish's heading could be determined for each midwater period. The presentation follows the convention that the part of the radius between the inner and outer circles corresponds to 100%. Total times in all sectors: period 1, 80 min; period 2, 31 min; period 3, 138 min; period 4, 127 min; period 5, 163 min.

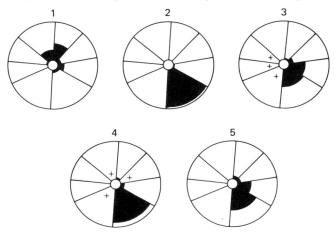

Fig. 12. Height above the bottom and compass headings during midwater period 1 which is broken into 4 separate parts between which the plaice returned to the bottom. The distribution of compass headings for each part is shown separately. Heights above the bottom of one metre or less were included in the category Just Above the Bottom and were not classified as midwater; OB indicates that the plaice was on the bottom. Total time in all sectors: part 1, 14 min; part 2, 16 min; part 3, 28 min; part 4, 22 min.

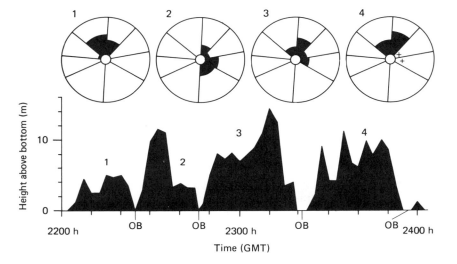

Fig. 13. Height above the bottom and compass headings during midwater period 4 which is broken into 4 separate sections between which the plaice returned to the bottom. The distribution of compass headings for each section is shown separately. OB indicates that the plaice was on the bottom; JOB, that it was just off the bottom. Total time in all sectors: part 1, 14 min; part 2, 104 min; part 3, 3 min; part 4, 6 min.

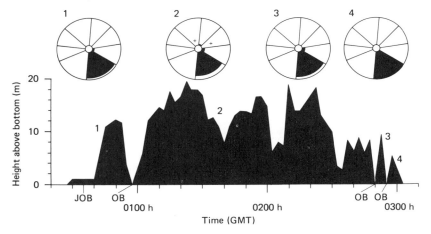

the fish moves into midwater; and secondly, how is the course subsequently maintained?

A lucky observation suggested a possible answer to the first question. Before one midwater period the plaice was seen to be swimming very close to the bottom along the crest of a sandwave before 'taking off'. When in midwater the fish maintained the same heading: a comparison with an aircraft was irresistible. This observation leads to the suggestion that the heading maintained in midwater was related to that observed shortly before 'take-off'. With this hypothesis in mind, the sector headings recorded during the hour or so before 'take-off' were broken down into 10-min sections and compared with the patterns characteristic of each of the corresponding five midwater periods. In each case the pattern recorded immediately before 'take-off' was similar to that observed in midwater. Two examples are shown in Figs. 14 and 15. During the first midwater period the headings of the fish were largely in sectors 6 and 7 (northwest to northeast). Fig. 14 shows that in the hour before 'take-off' the plaice turned clockwise from

Table 4. *Frequency with which a midwater plaice was recorded as maintaining its heading within one compass sector (about 45°) for a number of consecutive 1-min periods*

No. of consecutive 1-min periods	No. of observations
0–1	664
1–2	53
2–3	26
3–4	11
4–5	6
5–6	4
6–7	2
7–8	5
8–9	2
9–10	0
10–11	0
11–12	1
12–13	0
13–14	1

Note: Based on 775 observations, data from Alden chart record

heading southeast to head in a predominantly northwest-to-northerly direction in the final 10 min before moving into midwater. A similar change in heading was shown in the hour before midwater periods 2 and 3. But the fish was already heading in the midwater direction for some time before 'take-off' for periods 4 and 5 and here there were no changes: the data relating to midwater period 4 are shown in Fig. 15.

These observations are consistent with the hypothesis that the heading adopted in midwater is related to the heading of the fish on the bottom before 'take-off' and so, perhaps, to some feature of bottom topography. We think it is possible that sandwaves, whose crests are aligned at 90° to the tidal axis, could provide the bottom reference. The track of a plaice whose movements over the bottom appeared to be related to sandwaves has already been described (Greer Walker et al., 1978, p. 82).

While we have a reasonable but tentative suggestion for an answer to the first question, we have little to offer in respect of the second. The height of the fish above the bottom in turbid East Anglian waters on nights when neither moon nor stars were visible would seem to preclude the usual rheotropic clues or the use of celestial reference points. The way would appear to be left open for what would have once been dismissed as unlikely contenders: inertial guidance

Table 5. *Frequency with which a midwater plaice was recorded as maintaining its heading within two compass sectors (about 90°) for a number of consecutive 1-min periods*

No. of consecutive 1-min periods	No. of observations	No. of consecutive 1-min periods	No. of observations
0–1	16	12–13	3
1–2	7	13–14	1
2–3	6	14–15	3
3–4	7	15–16	0
4–5	5	16–17	2
5–6	2	17–18	1
6–7	2	18–19	1
7–8	1	19–20	0
8–9	3	20–21	2
9–10	0	28–29	1
10–11	4	37–38	1
11–12	0	57–58	1

Note: Based on 69 observations, data from Alden chart record

through the labyrinth or some on-line information of the earth's magnetic field through some unidentified sensor. We have given some thought to these matters. While our calculations suggest that the possibility of inertial guidance is not to be dismissed so lightly as it has been in the past, our conclusion is that the need is for more data, from both the sea and the laboratory, and less speculation.

Dr Arnold has summarized the lines of evidence which have led us to the conclusion that the plaice of the Southern Bight of the North Sea use tidal transport when on migration. Tidal transport has two potential advantages: it could save energy; and provided the fish catches and leaves the right tide at the appropriate moment, problems of direction do not arise. So we were surprised to find, in a fish living in shelf waters where tidal streams dominate the current regime, some evidence of an ability to maintain a heading when in midwater. However, these preliminary results are not inconsistent with Baker's (1978, p. 861) 'least navigation hypothesis': clearly we should give more thought to the design of

Fig. 14. A comparison between the distribution of compass headings recorded when the plaice was on the bottom with that recorded during midwater period 1. The plaice changed its heading from southeast (sectors 1 and 2) to northwest (sectors 5 and 6) before leaving the bottom at 2207 h. When in midwater the compass headings lay between northwest and northeast (sectors 6 and 7).

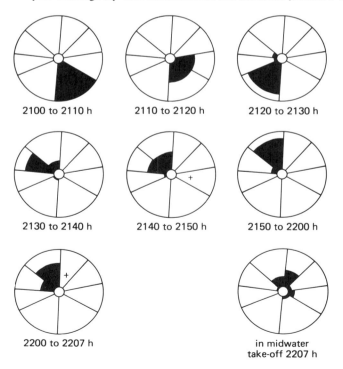

2100 to 2110 h   2110 to 2120 h   2120 to 2130 h

2130 to 2140 h   2140 to 2150 h   2150 to 2200 h

2200 to 2207 h

in midwater
take-off 2207 h

field and laboratory experiments to see if a similar faculty is evident in other species of fish.

Fig. 15. A comparison between the distribution of compass headings recorded when the plaice was on the bottom with that recorded during midwater period 4. The plaice maintained a southeasterly heading when on the bottom and after moving into midwater at 0040 h.

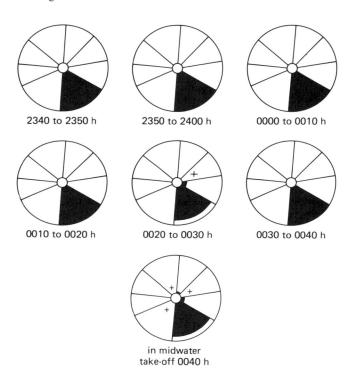

2340 to 2350 h   2350 to 2400 h   0000 to 0010 h

0010 to 0020 h   0020 to 0030 h   0030 to 0040 h

in midwater
take-off 0040 h

References

Anon. (1960). Proc. Joint Scient. meeting of ICNAF, ICES, and FAO on fishing effort, the effect of fishing on resources and the selectivity of fishing gear. I: – reports. *Special Publications of the International Commission for the Northwest Atlantic Fisheries,* (2), 45 pp.

Baker, R. R. (1978). *The Evolutionary Ecology of Animal Migration.* London: Hodder & Stoughton. 1012 pp.

Bokuniewicz, H. (1980). Groundwater seepage into Great South Bay, New York. *Estuarine & Coastal Marine Science,* **10,** 437–44.

Boyd Dawkins, W. (1898). On the relation of geology to engineering. *Minutes of the Proceedings of the Institution of Civil Engineers,* **134,** 254–77.

Brereton, N. R. & Downing, R. A. (1975). Some applications of thermal infra-red linescan in water resources and studies. *Water Services,* **79,** 91–6.

Buljan, M. & Zore-Amanda, M. (1976). Oceanographical properties of the Adriatic Sea. *Oceanography and Marine Biology Annual Reviews,* **14,** 11–98.

Carew, R. (1811). *Carew's Survey of Cornwall;* to which are added notes illustrative of its history and antiquities, by the late Thomas Tonkin, Esq. London: Faulder. 459 pp.

Cohen, D. M. (1970). How many recent fish are there? *Proceedings of the California Academy of Sciences,* **38,** 341–5.

Cushing, D. H. (1969). Migration and abundance. In *Perspectives in Fisheries Oceanography,* pp. 207–12. *Bulletin of the Japanese Society of Fisheries and Oceanography* (special number).

Cushing, D. H. (1975). *Marine Ecology and Fisheries.* Cambridge: Cambridge University Press. 278 pp.

Davies, M. C. (1973). A thermal infra-red linescan survey along the Sussex coast. *Water and Water Engineering,* **77,** 392–6.

Emery, K. O. & Hulsemann, J. (1963). Submarine canyons of Southern California. Part 1. Topography, water, and sediments. *Allan Hancock Pacific Expedition,* **27** (1), 80 pp.

FAO (1978). *Yearbook of Fisheries Statistics,* **44,** *Catches and landings,* 1977. 343 pp.

Favorite, F., Dodimead, A. J. & Nasu, F. (1976). Oceanography of the subarctic Pacific region, 1960–71. *Bulletin of the International North Pacific Fisheries Commission,* **33,** 187 pp.

Foerster, R. E. (1937). The return from the sea of sockeye salmon (*Oncorhynchus nerka*) with special reference to percentage survival, sex proportions and progress. *Journal of the Biological Board of Canada,* **3,** 26–42.

Forster, G. R. (1974). The ecology of *Latimeria chalumnae* Smith: results of field studies from Grande Comore. *Proceedings of the Royal Society of London,* B, **186,** 291–6.

Gerking, S. D. (1959). The restricted movements of fish populations. *Biological Reviews,* **34,** 221–42.

Greer Walker, M., Harden Jones, F. R. & Arnold, G. P. (1978). The movements of plaice (*Pleuronectes platessa* L.) tracked in the open sea. *Journal du Conseil. Conseil International pour l'Exploration de la Mer,* **38,** 58–86.

Hara, T. J. & Brown, S. B. (1979). Olfactory bulbar electrical responses of rainbow trout (*Salmo gairdneri*) exposed to morpholine during smoltification. *Journal of the Fisheries Research Board of Canada,* **36,** 1186–90.

Harden Jones, F. R. (1980). The migration of plaice (*Pleuronectes pla-*

tessa) in relation to the environment. In *Fish Behaviour and Its Use in the Capture and Culture of Fishes*, ed. J. E. Bardach, J. J. Magnuson, R. C. May & J. M. Reinhart, pp. 383–99. Manila: International Center for Living Aquatic Resources Management.

Hasler, A. D. & Scholz, A. T. (1980). Artificial imprinting: a procedure for conserving salmon stocks. In *Fish Behaviour and Its Use in the Capture and Culture of Fishes*, ed. J. E. Bardack, J. J. Magnuson, R. C. May & J. M. Reinhart, pp. 179–99. Manila: International Center for Living Aquatic Resources Management.

Hughes, F. W., Coachman, L. K. & Aagaard, K. (1974). Circulation, transport and water exchange in the western Bering Sea. In *Oceanography of the Bering Sea*, D. W. Hood & E. J. Kelley, pp. 59–98. In *Institute of Marine Sciences Occasional Publications* (2). Fairbanks, Alaska: University of Alaska.

Huntsman, A. G. (1934). Herring and water movements. In *James Johnstone Memorial Volume*, pp. 81–96. Liverpool: Liverpool University Press, Lancashire Sea-Fisheries Laboratory.

Huxley, T. H. (1893). The progress of science. In *Collected Essays, 1: Method and Results*, pp. 42–129. London: Macmillan.

Huxley, T. H. (1894). The problems of the deep sea. In *Collected Essays, 8*, pp. 37–68. London: Macmillan.

Idrac, P. (1933). Appareil Idrac pour la mesure de courants verticaux sousmarins. *Bulletin de l'Institut Océanographique*, (637), 3 pp.

Kinder, T. H., Coachman, L. K. & Galt, J. A. (1975). The Bering Slope Current System. *Journal of Physical Oceanography*, **5**, 231–44.

Kohout, F. A. (1966). Submarine springs: a neglected phenomenon of coastal hydrology. In *Symposium on Hydrology and Water Resources Development, Ankara*, pp. 391–413. Central Treaty Organisation.

Le Danois, E. (1934). Les transgressions océaniques. *Revue des Travaux de l'Office de Pêches Maritime*, **7**, 369–459.

Lowndes, A. G. (1941). The displacement method of weighing living aquatic organisms. *Journal of the Marine Biological Association of the United Kingdom*, **25**, 555–574.

Mailyan, R. A. (1970). Factors limiting the abundances of fishes. *Journal of Ichthyology* **10**, 394–99.

Manheim, F. T. (1967). Evidence for submarine discharge of water on the Atlantic continental slope of the Southern United States and suggestions for further search. *Transactions of the New York Academy of Sciences*, Ser. 2, **29**, 839–53.

Marshall, N. B. (1964). Fish. In *Antarctic Research*, ed. R. Priestly, R. J. Adie & G. de Q. Robin, pp. 206–18. London: Butterworth.

Mitson, R. B., Storeton-West, T. R. & Pearson, N. D. (in preparation). An acoustic transponding compass tag.

Nikolsky, G. V. (1963). *The Ecology of Fishes*. New York: Academic Press. 352 pp.

Runnström, S. (1941a). Racial analysis of the herring in Norwegian waters. *Fiskeridirektoratets Skrifter, Serie Havundersøkelser*, **6**(7). 110 pp.

Runnström, S. (1941b). Quantitative investigations on herring spawning and its yearly fluctuations at the west coast of Norway. *Fiskeridirektoratets Skrifter, Serie Havundersøkelser*, **6**(8). 71 pp.

Sars, G. O. (1879). Report of practical scientific investigations of the cod fisheries near the Loffoden Islands, made during the years 1864–69. *Report of the United States Commissioner of Fisheries for 1877*, pp. 565–611.

Serobaba, I. I. (1968). Spawning of the Alaska pollack *Theragra chalco-gramma* (Pallas) in the Northeastern Bering Sea. *Problems in Ichthyology* **8,** 789–98.

Serobaba, I. I. (1970). Distribution of walleye pollack *Theragra chalco-gramma* (Pallas) in the Eastern Bering Sea and prospects of its fishery. *Trudy VNIRO,* **70:** *Isvestiya TINRO,* **72,** 433–42, 442–51. In *Soviet Fisheries Investigations of the Northeastern Pacific,* part V, ed. P. A. Moiseev, IPST 60044 **6** (1972), 462 pp.

Smitt, F. A. (1895). *A History of Scandinavian Fishes,* part 2, 2nd edn. Stockholm: Norstedt. 1240 pp.

Storer, J., (1815). On an ebbing and flowing stream discovered by boring in the harbour of Bridlington. *Philosophical Transactions of the Royal Society,* 54–9.

Svetovidov, A. N. (1970). Some comments on R. A. Mailyan's paper "Factors limiting the abundance of fishes". *Journal of Ichthyology* **10,** 399–403.

Tibbetts, G. R. (1971). *Arab Navigation in the Indian Ocean before the Coming of the Portuguese.* Oriental Translation Fund, NS, **42.** 614 pp.

Versey, H. C. (1946). The Humber Gap. *Transactions of the Leeds Geological Association,* **6,** 6–9.

RAYMOND J.O'CONNOR

Comparisons between migrant and non-migrant birds in Britain

Introduction

Ornithologists have long considered the questions raised by the phenomenon of bird migration (Matthews, 1955). Much of this attention has focussed on the immediate or proximate factors behind migration (Emlen, 1975), with considerably less attention given to the ultimate factors behind the behaviour (Baker, 1978). One can obviously speculate about the ultimate causation of migration but it is a topic not particularly amenable to experimental or comparative studies: convincing proof of the survival value of migration essentially requires comparison of the survival of individuals which migrate with the survival of comparable individuals which do not. In general, one cannot obtain comparable samples of birds from the two classes: either a species migrates completely, so that a population of non-migrant birds of the same species is not available for study or, where the species concerned is a partial migrant, those individuals who do migrate are generally of a different sex or age class from those that do not (Lack, 1943). Against this pattern of known variation amongst partial migrants it is difficult to be confident that migrating and non-migrating individuals are similar except in their migratory trait and therefore difficult to be confident that one is assessing the survival value of migration rather than of some more basic but unknown difference between the two groups. To assess the selective value of migration by direct means is thus problem-ridden.

An alternative approach to this question is to compare the characteristics of migrant species with those of non-migrant species. Life-history characteristics are evolved traits, subject to the constraints of the evolutionary history of the species involved. It may therefore be possible to gain an insight into the evolution of migration by comparing the life-history characteristics of migrants with those of non-migrants. Any general differences between the two groups – migrants and residents – are likely also to reflect the influence of those factors leading to the evolution of the migratory habits in the first place. Any such differences therefore provide circumstantial evidence as to the ultimate factors behind migration.

The present paper outlines a first attempt at applying this approach to the study of migration. Much basic information about the breeding biology and population behaviour of farmland birds in Britain is available, largely as a result of the systematic collection of data by the British Trust for Ornithology (BTO). By examining this material in the cases of eleven migrant species and thirty-one non-migrant species, all of them breeding relatively commonly on farmland in Britain, an initial assessment of the possibility of detecting differences between the two groups in their life-history ecologies could be made. For the present analysis the species groups were simplified by regarding the birds as either total migrants – those species the vast majority of whose individuals spend the winter outside the British Isles – or resident species – those species in which a significant proportion of the population winters within the British Isles. This simplification means that a considerable number of partial migrants (in the sense of Lack (1943)) were treated as sedentary. It also ignores recent reports of small but growing numbers of certain migrant species, notably the blackcap *Sylvia atricapilla,* overwintering in Britain or Ireland. As part of the same simplification for this first analysis, no account was taken of regional differences in the variables studied.

Materials and methods

Materials were collated with regard to five aspects of each species' biology: (1) biometrics, (2) population changes, (3) survival, (4) reproductive effort, and (5) distribution and abundance. In addition, for resident species only, the population parameters were extended to involve measurements of the recovery rate of each species population level following a catastrophic perturbation imposed upon it by the chance succession of two extremely severe winters.

Biometric variables were included because body size is known to be an important parameter of life-history biology (Calder, 1974; Southwood, 1976). In the present study body weights were extracted from an unpublished compilation by S. Cousins, with additional material from the files of the BTO Ringing Scheme. In some species body weight is specific but in others there are significant differences between males and females (Selander, 1966). To allow for this in the present study wing length was taken as an approximate indicator of sexual dimorphism, data being obtained in the first instance from Witherby, Jourdain, Ticehurst & Tucker (1938) but updated from BTO ringing records. As an index of the extent of dimorphism within each species I used the ratio of mean male wing length to mean female wing length to calculate a percentage dimorphism index.

Changes in population level were extracted from the records of the Common Birds Census (CBC) Scheme for farmland. This scheme has been organized by the BTO since 1961 and provides an annual index of percentage change of bird populations on a sample of farms throughout Britain (numbering about one hundred each year). Fig. 1 shows the results obtained for the blackbird *Turdus merula,* the commonest bird on farmland in Britain.

The CBC technique measures percentage changes from year to year more accurately than absolute densities and for this reason an arbitrary index level of 100 in the year 1966 is used as the scaling factor for the index. Changes in the index value then reflect year-to-year changes in the population level of the species concerned. The various possibilities of bias in this index have been reviewed in a number of methodological papers (e.g. Williamson & Homes, 1964; Taylor, 1965; Bailey, 1967), to ensure that the index is free from the effects of all likely biasses. For most farmland species CBC index data are available from 1962, thus providing a suite of comparable population data for these species. However, the first two index values were somewhat depressed by the occurrence of a rather cold winter in 1961–2 and by an excessively severe winter in 1962–3, so that the populations of these species were very severely reduced from their levels before these winters (Dobinson & Richards, 1964). Fig. 1 shows this effect clearly, displaying the recovery phase for blackbird populations until about 1966, after which the population showed relatively restricted fluctuations about a fairly stable equilibrium level. Although cold winters are a recurring feature of the environment of resident species in Britain they occur at relatively long intervals (the previous winter as severe as 1962–3 was 1947–8). For this reason I

Fig. 1. The Common Birds Census index for blackbirds *Turdus merula* counted on farmland in Britain. Note the population decrease following the 1962–3 winter, the subsequent recovery, and the relatively stable population density since then. The 1966 density was arbitrarily equated to an index value of 100 (Bailey, 1967).

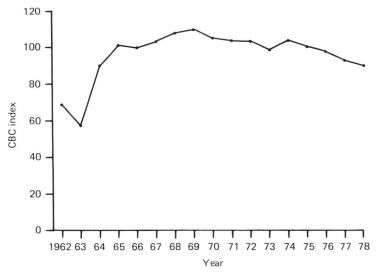

chose to restrict the analysis of population changes for migrants and residents of the CBC data for the period 1966–77, thereby excluding the worst effects of the 1962–3 winter upon the resident species. Migrant species would not have been affected by this winter, of course, so comparability of the 'regular' environments for the two species required the exclusion of such catastrophic events.

In the present study three measures of population changes were analysed. The first was the frequency with which the population rose or fell between years. As a measure of this the numbers of population decreases between successive pairs of breeding seasons within the period 1966–77 were tabulated. With a total of ten year-to-year changes available within this period the expected value for the count of population decreases in a population at equilibrium was five and the expected value for the number of population increases was also five. In fact, in a very small number of cases the population level did not change between years and in such cases the counts of decreases and of increases were each incremented by 0.5. Thus, significant departure of the distribution of population decreases from five would indicate a significant tendency for the population either to increase or to decrease over this eleven-year period.

The amplitude of population changes was also indexed for the present study. Southwood's (1976, 1977) reviews of animal bionomic strategies showed that such strategies could result in asymmetric population changes. That is, it is conceivable that a population might more readily decrease than recover from that decrease. For this reason it was desirable to segregate the size of population increase from the size of population decrease in comparing resident and migrant strategies. The population level graphs such as Fig. 1 were scanned and the positive year-to-year changes extracted and their average value calculated to yield an *increase percentage*. This thus referred to the size of the population change between two successive years given that the population was in fact increasing. Similarly, the data were examined for year-to-year decreases and the average size of such decreases calculated to give an index of the population decrease when it was in fact decreasing. These two measures thus provided for control of the possibility of asymmetry of population change.

In the course of the analysis it emerged that it was desirable to assess for the resident species populations their rates of recovery from the effects of the average 1962–3 winter. As a measure of this recovery rate a *resilience index* was used. This resilience index was derived for each species as the percentage change in that species' population size between the summers of 1963 and 1964. Since most bird populations in Britain were enormously reduced by the experience of the severe winter most farmland populations were increasing from very low densities, so the observed population increase would have been something fairly close to the 'intrinsic rate of increase' of a species expanding in an ecological vacuum (Krebs, 1972). Note, though, that although the resilience index used here is very close in concept to the idea of a short 'return time' used by theoretical ecologists, e.g. Southwood, May, Hassell & Conway (1974), it is not in fact identical to that idea.

Survival

An index of survival for each species was calculated from the five-year recovery cohorts reported in the BTO Ringing Scheme Annual Report for 1976 (Spencer & Hudson, 1978) for the years 1972–6. Survival was indexed for each species as an exponential rate of decay for a recovery cohort of known ringing dates (C. J. Mead & D. Sales in preparation). The index thus calculated was not an absolute measure of survivorship as used by Haldane (1955) but was adequate for the comparative purpose indicated here.

Reproductive effort

As a measure of reproductive effort the seasonal egg productivity of each species was calculated by multiplying clutch size by the number of successful clutches laid in each season, using data tabulated by Harrison (1975) and Campbell & Ferguson-Lees (1972). The two separate components of the productivity measure, clutch size and clutches per season, were both retained as individual variables for examination within the analysis.

Abundance and distribution

Sharrock (1976) reports on the results of a five-year survey of the distribution of breeding birds in Britain. Each map in that *Atlas* shows the presence or absence of a species within each 10-km square of the Ordnance Survey grid of the British Isles. Consequently, one index of the distribution of each species within Britain is the number of 10-km squares in which the species was recorded at some time in the course of this five-year period. The methodology of the *Atlas* was such that species which were very abundant within a 10-km square were more likely to be detected than were species which were less abundant within this square. It follows that the number of squares with a species present, hereafter referred to as the *ubiquity* of the species, was monotonically related to the abundance of the species.

A second measure of the abundance of each species was obtained from the number of recorded territories for the species in the farmland Common Birds Census baseline year of 1966 (Bailey, 1967). In 1965–6 a total of 19 826 acres of farmland was counted as part of the CBC scheme. With comparable effort of recording for each species on the farms the number of recorded territories thus provided a direct measure of the absolute abundance of the species in that year. As already noted, many of the species had already significantly recovered from the effects of the 1962–3 winter by 1966. Thus, the 1966 counts of territories yield an index of the absolute densities of the various species on British farmland in 1966, thereby providing a direct measure of abundance to supplement that inferred from the ubiquity measure derived from the *Atlas*.

Species studied

Table 1 lists the species whose data were analysed in the course of the present study. Three species, red-legged partridge *Alectoris rufa*, grey partridge *Perdix perdix*, and pheasant *Phasianus colchicus*, were omitted from certain analyses in the absence of adequate survival data.

Results

Abundance

Are migrants more or less numerous on farmland in Britain than are resident species? Two measures of 'abundance' were considered in addressing this question. The first was *ubiquity*, the number of 10-km OS squares recorded as containing breeding individuals of each species during the 1968–72 *Atlas of*

Table 1. *Species whose data were analysed in the present study*

Residents	
Mallard *Anas platyrhynchos*	Song thrush *Turdus ericetorum*
Red-legged partridge *Alectoris rufa*	Blackbird *Turdus merula*
Partridge *Perdix perdix*	Robin *Erithacus rubecula*
Pheasant *Phasianus colchicus*	Dunnock *Prunella modularis*
Moorhen *Gallinula chloropus*	Pied wagtail *Motacilla alba*
Lapwing *Vanellus vanellus*	Starling *Sturnus vulgaris*
Skylark *Alauda arvensis*	Greenfinch *Chloris chloris*
Carrion crow *Corvus corone*	Goldfinch *Carduelis carduelis*
Jackdaw *Corvus monedula*	Linnet *Carduelis cannabina*
Magpie *Pica pica*	Bullfinch *Pyrrhula pyrrhula*
Great tit *Parus major*	Chaffinch *Fringilla coelebs*
Blue tit *Parus caeruleus*	Yellowhammer *Emberiza citrinella*
Long-tailed tit *Aegithalos caudatus*	Corn bunting *Emberiza calandra*
Tree creeper *Certhia familiaris*	Reed bunting *Emberiza schoeniclus*
Wren *Troglodytes troglodytes*	Tree sparrow *Passer montanus*
Mistle thrush *Turdus viscivorous*	
Migrants	
Turtle dove *Streptopelia turtur*	Garden warbler *Sylvia borin*
Cuckoo *Cuculus canorus*	Whitethroat *Sylvia communis*
Swallow *Hirundo rustica*	Lesser whitethroat *Sylvia curruca*
Sedge warbler *Acrocephalus schoeno-baenus*	Willow warbler *Phylloscopus trochilus*
	Chiffchaff *Phylloscopus collybita*
Blackcap *Sylvia atricapilla*	Spotted flycatcher *Muscicapa striata*

Breeding Birds (Sharrock, 1976). In principle such a measure is not necessarily correlated with numerical abundance, since the species might be widely distributed at low densities. However, the survey methods used for the *Atlas* tended to favour discovering the more abundant species. Use of the *Atlas* data had the advantage in assessing abundance of integrating field data from a five-year period (1968–72), thus avoiding the effects of vagaries of weather and migration conditions in any individual year. The second measure of abundance analysed – the densities of birds recorded on farmland in the Common Birds Census in 1966 (the standard reference year for subsequent CBC index calculations) – is potentially subject to bias from these sources but provides a direct measure of avian densities in a single year, for comparison with the ubiquity results and without the biasses of the former. The two measures were in fact mutually correlated for all 42 species examined ($r = 0.390$; $P < 0.02$) and this correlation did not differ significantly between the residents and the migrants ($r = 0.376$; $P < 0.05$; and $r = 0.354$; not significant (n.s.), respectively). The relationship is not linear, however, since most species (but especially residents) are virtually ubiquitous above a threshold density, and this aspect has not been taken into account here.

Table 2 shows the distribution of ubiquities for residents and migrants. On average, resident species were present in 3172 *Atlas* squares whilst migrants were present in 2681 squares. The distributions for both species groups were highly skewed, so median statistics were more appropriate. For the residents the median was 3380 whilst for migrants it was 2929 but the difference between the two distributions only just approached statistical significance (Fisher exact probability = 0.079, Median test (Siegel, 1956)).

Table 2. *Ubiquity distributions for resident and migrant species in Britain*

Ubiquity interval (Atlas registrations)[a]	No. of species in stated interval	
	Residents	Migrants
0–1000	1	0
1001–1500	1	2
1501–2000	1	1
2001–2500	1	1
2501–3000	2	2
3001–3500	14	2
3500–4000	11	3
Mean ubiquity ± s.d.[b]	3172 ± 676	2681 ± 917

[a] From Sharrock (1976).
[b] With a combined median of 3340 *Atlas*-square registrations a Median test (Siegel, 1956) gave a Fisher exact probability of $P = 0.079$ for a difference between the resident and migrant distributions.

Table 3 presents the density distribution for the two species groups in 1966. For the resident species the average number of clusters recorded was 513, corresponding to density on the census plots of 6.38 pairs/km² whilst for the migrant species the average number of clusters recorded was 228, corresponding to absolute densities of 2.84 pairs/km². Thus residents were on average more than twice as numerous as were migrants in 1966. The distributions were again highly skewed, particularly for the resident species: of the 15 species for which more than 500 territories were counted in 1966, no less than 13 were residents: only willow warbler *Phylloscopus trochilus* and whitethroat *Sylvia communis* amongst the migrants attained this population level. Overall, migrant species accounted for only 13.6% of the territories counted although the 11 migrant species formed 26.2% of the 42 species considered here ($\chi^2 = 4.59$, $P < 0.05$). Thus, the migrants accounted for significantly fewer of the birds present on farmland than one would expect from the number of species involved. This is supported by weaker statistical evidence for a difference in the two distributions of Table 3 (Fisher exact probability = 0.079, Median test).

Summarizing this section, both ubiquity and densities suggest that individual migrant species were less widespread and/or numerous than were individual resident species and the migrant species collectively accounted for disproportionately few of the breeding pairs found on farmland in Britain.

Population changes

Do migrants differ from residents in the behaviour of their populations from year to year? This question was examined on the basis of the Common Birds Census indices for the period 1966–77.

Table 4 examines the frequency with which year-on-year population declines were observed amongst the migrant and amongst the resident species. Of the 23 species showing 4 or fewer falls in population level over the 11-year period, no

Table 3. *Contingency table for species with low (below median of 180) and high (above median of 180) census counts of territories in the Common Birds Census in the base-line year of 1966, based on 8030 ha of farmland counted in 1965 and 1966*

	No. of species	
Abundance	Residents	Migrants
Fewer than 180 territories	13	8
More than 180 territories	18	3
	Fisher exact probability = 0.079	

less than 21 were residents whilst 9 of the 19 species showing 5 or more falls over the period were migrants; the difference between the two groups is highly significant (Fisher exact probability $= 0.006$). Migrant species were thus more susceptible to population decrease.

Fig. 2 examines the size of such decreases as occurred and shows that not only did the residents experience population falls less frequently than migrants but that the falls were also smaller than in migrants. No less than 9 of the 11 migrant species experienced annual decreases worse than the median value of 8.7% whilst only 10 of the 31 resident species experienced decreases as severe ($\chi^2 = 6.17$, $P < 0.02$). On the other hand, Fig. 3 shows that migrants achieved bigger population increases from year to year than did the residents: only 12 of the 31 residents exceeded the median year-on-year increase value of 10.55% whilst 9 of the 11 migrant species did so ($\chi^2 = 4.43$, $P < 0.05$). Thus, although migrants experienced relatively larger decreases than did the resident species, they compensated by achieving larger increases in other years.

Fig. 4 examines the extent to which this density compensation occurs intra-specifically. In this figure the 'recovery' percentage is that needed to reverse the effect of the population decrease experienced to maintain comparability with the 'increase' figures. For example, if a species index dropped from 100 to 80 between years, the surviving birds need a 25% (not a 20%) population increase to compensate for the decline. The figure shows that the majority of species tended to cluster around the line of equal increases and recoveries, with a slight

Table 4. *Frequency distributions for residents and migrants of year-on-year population decrease[a] between 1966 and 1977*

Frequency of decrease between successive years	No. of species	
	Residents	Migrants
2	4	0
3	3	1
4	14	1
5	2	4
6	6	2
7	1	3
8	1	0
	Fisher exact probability[b] $= 0.006$	

[a]As measured by the farmland Common Birds Census index (Bailey, 1967).
[b]For categories 2–4 pooled against 5–8 pooled and tested for resident–migrant difference.

bias towards net increase. Red-legged partridge was the species most prone to net increase, achieving an average increase of more than 35% when increasing, against the need for a recovery of less than 10% when in decline. The long-tailed tit *Aegithalos caudatus* was the species most prone to decrease during the 11-year period considered but this was largely due to a decrease of over 45% experienced in a single year (that between summer 1975 and summer 1976). The

Fig. 2. Distributions of population decrease amplitudes for residents and migrants. For each species a mean decrease value was computed by averaging all year-on-year percentage decreases shown by the farmland Common Birds Census indices over the period 1966–77. The distributions for residents (upper) and migrants (lower figure) differ significantly (Median test $X^2 = 6.17$, $P < 0.02$).

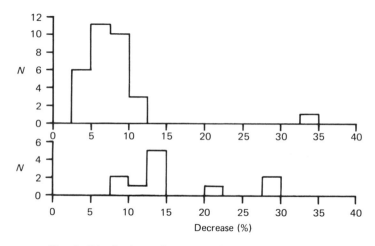

Fig. 3. Distributions of mean population increases undergone by resident (upper) and by migrant (lower figure) species in years of increase. Details as for Fig. 2. The distributions differ significantly (Median test $X^2 = 4.43$, $P < 0.05$).

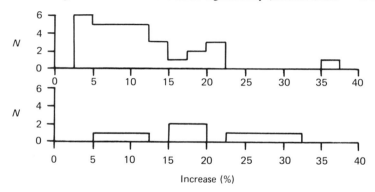

whitethroat also shows up badly on this diagram, with a year-on-year recovery value of 40% against an average population gain in years of increase of less than 20% but this was largely due to the effect of the Sahelian drought which destroyed areas of the winter habitat of the species during the mid-60s (Winstanley, Spencer & Williamson, 1974). Overall, the increase and recovery values for individual species are only weakly correlated in migrants ($r = 0.512, P = 0.109$) and in residents ($r = 0.303; 0.05 < P < 0.10$) but when the anomalous species are excluded the remaining species show statistically significant correlations within both groups. For the migrants, exclusion of the whitethroat yields a correlation between the increase and recovery percentage of 0.663 ($P < 0.05$) whilst within the resident species exclusion of the long-tailed tit yielded a correlation of 0.538 ($P < 0.002$). Inspection of Fig. 4 shows that a small majority of the data points lie above the line (suggesting a net tendency to increase) but this is statistically significant neither in the combined data (Signs test, 2-tailed $P = 0.118$) nor in the case of the resident species alone (2-tailed $P = 0.073$). The evidence thus suggests that most species experience population declines between years no larger than they can compensate for by an increase in a year of rising populations.

Fig. 4. Relationship between average annual increases and average annual 'recovery' percentages (see text) for resident (filled circles) and migrant (hollow circles) species when years of increase are segregated from years of decrease during the period 1966–77. Changes are those shown by the Common Birds Census indices for farmland and are expressed as percentages of the *lower* index in each case. RL, red-legged partridge; WH, whitethroat; LTT, long-tailed tit.

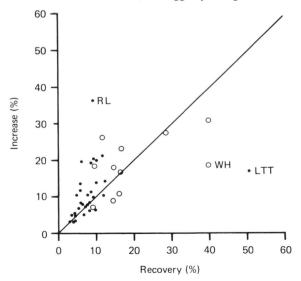

Reference back to Table 4 shows that for migrants the median number of year-on-year population declines lay between five and six over an 11-year period. If the population changes experienced by the migrants were due to random factors one would expect about half the annual changes to involve falls in the CBC index and about half rises, as in fact appears to be the case for the migrants. With the evidence of Fig. 4 this suggests that the migrant populations were approximately in equilibrium over this period. For resident species, on the other hand, the significantly lower frequency of population decreases revealed by Table 4 coupled with the correlations between increases and decreases revealed by Fig. 4 implies that the resident species were increasing, on average, over the same period. Since the achieved increases were equal in magnitude to those necessary to offset such decreases as occurred, this finding implies that the birds were experiencing conditions favourable to increase and not displaying a differential intrinsic rate of increase.

Survival

Fig. 5 shows the distribution of survivorship for residents and for migrants. Median survival of all 42 species was 0.482 and there was a slight bias towards the residents surviving better than the migrants (18 of 31 residents exceeded the median in survival against 3 of 11 migrant species) but the difference was well short of statistical significance ($\chi^2 = 1.97$, $0.20 > P > 0.10$). However, it is well known that survival in a wide range of animals is correlated with body weight and there are in fact major differences between the weight distributions of migrants and of residents (Fig. 6). The median weight for the 42 species considered here was 21.7 g and this weight was exceeded by only 2 of the 13 migrants but by 18 of 27 resident species, a highly significant difference

Fig. 5. Survival rate index distributions for resident (upper) and migrant (lower figure) species. The distributions do not differ significantly (Median test $\chi^2 = 1.97$, n.s.).

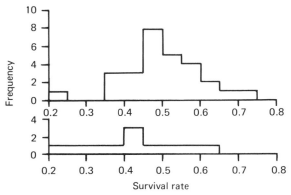

(Median test, $P < 0.003$). Within the residents, survival was significantly correlated with body weight measured on a logarithmic scale ($r = 0.381$, $P < 0.05$) but within the migrants the correlation was not significant ($r = 0.059$, n.s.). Comparison of the regression lines for the two groups showed that the migrants were significantly more variable in body weight than were resident species (F $(9,27) = 5.03$, $P < 0.01$). Fig. 6 shows that the general trends within the two groups are obscured by two individual species, the migrant cuckoo and the resident corn bunting. One might expect the cuckoo to be anomalous in such a plot since its habits of parasitizing other birds' broods might expose it to less risk of mortality during the breeding season. For the corn bunting there also exists an *a priori* case for expecting an anomaly, the species being polygynous. Omission of these two species does not, however, influence significantly the overall conclusions reached above.

Sexual dimorphism in body size potentially reflects differential mortality between the two sexes and might also differ between residents and migrants. Table 5 shows the extent of the dimorphism in resident and in migrant species and shows that there is only a slight bias towards greater dimorphism on the part of the resident species (Kolmogorov–Smirnov test, $\chi^2 = 1.04$, n.s. (Siegel, 1956)). Thus the absolute incidence of sexual dimorphism within the two groups hardly differs. Fig. 7, on the other hand, shows that there were significant differences in the relationship of survival to dimorphism within the migrants and

Fig. 6. Survival rate in relation to body weight in residents and in migrants. Regression lines fitted separately to residents (filled circles) and migrants (hollow circles) differ significantly in variance. CK, cuckoo; CB, corn bunting.

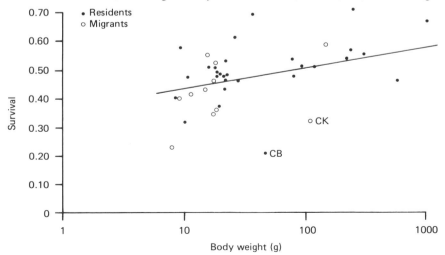

Table 5. *Frequency distributions for resident and migrant species of sexual dimorphism in wing length*

| Dimorphism interval (%)[a] | No. of species | |
	Residents	Migrants
0.0–2.0	7	4
2.1–4.0	10	4
4.1–6.0	7	2
6.1–8.0	4	1
8.1–10.0	1	0
10.1–12.0	0	0
12.1–14.0	2	0
	Kolmogorov-Smirnov $\chi^2 = 1.04$, n.s.[b]	

[a] Percentage deviation of mean male wing length from mean for females.
[b] From Siegel (1956).

Fig. 7. Relationships between survival index and the extent of sex dimorphism in wing length (male–female difference as percentage of female). With untransformed variables, neither regression is statistically significant. Upper figure, residents; lower figure, migrants.

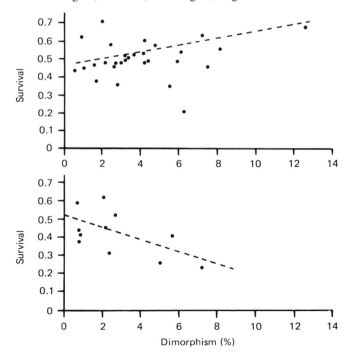

within the residents. Amongst migrants, survivorship is negatively correlated with percentage dimorphism between the two sexes ($r = -0.576$; $0.10 > P > 0.5$) whilst within the residents there is a weak positive correlation between the two variables ($r = 0.194$). Although neither regression is statistically significant, the difference between the two groups is ($F(1,36) = 7.98$, $P < 0.05$). That is, the more sexually dimorphic residents survive better than do the more sexually dimorphic migrants but the less dimorphic residents are very similar in survival to the least dimorphic of the migrants.

Fig. 8 examines the relationship between survival and ubiquity. Amongst the resident species, those with high survival tended to be more widespread ($r = 0.384$, $P < 0.05$), whilst amongst the migrants a relationship in the opposite direction was suggested ($r = -0.448$, n.s.). The two relationships differed significantly in slope (F $(1,26) = 6.77$, $P < 0.05$); the most widespread residents

Fig. 8. Ubiquity (number of *Atlas* 10-km squares) in relation to survival index in residents (upper figure) and in migrants (lower figure). For residents the regression line is statistically significant ($P < 0.05$): $y = 2287.9 + 2002.5\,x$ with $r = 0.384$ ($P < 0.05$). For migrants, ubiquity was independent of survival ($r = 0.448$, n.s.).

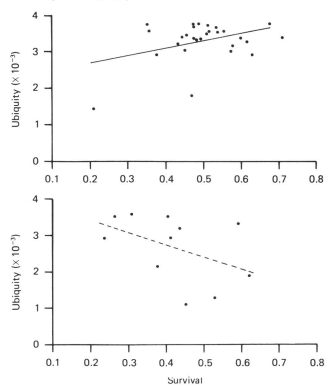

were those with high species survival whilst the most widespread migrants were those with relatively poor survival.

The relationship of species survival to the temporal behaviour of the population was examined by calculating regressions of the frequency of year-on-year population decline on survival. For the resident species the regression equation obtained was

$$\text{frequency} = 5.82 - 3.30\,S, \qquad r = -0.237, \text{n.s.}$$

where S is the survival index for the species, whilst amongst the migrants the regression obtained was

$$\text{frequency} = 4.42 + 2.46\,S, \qquad r = 0.237, \text{n.s.},$$

showing that in neither case was there any significant relationship. Despite the difference in the signs of the two slopes their difference was not statistically significant ($F\,(1,35) = 2.08$, n.s.) The frequency of population declines between years thus did not depend on survival either in residents or in migrants. The effect of survival on the size of population changes was similarly calculated by regression analysis, yielding the equations

$$\text{increase} = 12.63 - 4.89\,S, \qquad r = -0.094, \text{n.s.}$$

whilst amongst the migrants the regression equation obtained was

$$\text{increase} = 17.74 + 2.04\,S \qquad r = 0.032, \text{n.s.},$$

where the increase is expressed as percentage change in each case. Again, therefore, survivorship did not influence the dynamics of the populations of residents or of migrants.

Egg production

Fig. 9 sets out the frequency distribution of egg production (number of eggs produced per season) for residents and migrants and shows that there is little difference between the two groups of birds ($\chi^2 = 0.49$, n.s.). Egg productivity can be partitioned for each species into the separate contributions of clutch size and of clutches produced per season. Table 6 shows that there was no difference between the distribution of clutch sizes produced by the two groups ($\chi^2 = 0.18$, n.s.) and Table 7 shows that both groups have similar distributions of clutches produced per year ($\chi^2 = 1.49$, n.s.). Fig. 10 examines the relationship between ubiquity and egg productivity within the two groups. Within the residents ubiquity was weakly negatively correlated with productivity ($r = -0.218$, n.s.) but within the migrants the two variables were strongly correlated ($r = 0.707$, $P < 0.02$). The two regressions of course differed significantly ($F(1,38) = 15.30$, $P < 0.01$). This result implies that the resident species producing most eggs each year were relatively restricted in distribution whilst amongst the migrants the most productive species were also the most widespread. This finding contrasts strongly with that of Fig. 8.

As with survival, it was possible to examine how egg productivity influenced the temporal behaviour of the populations of residents and migrants. Calculation of regression equations for the frequency of year-on-year population declines gives for resident species

$$\text{frequency} = 3.41 + 0.08\,Pr \qquad r = 0.181, \text{ n.s.},$$

Table 6. *The distribution of clutch sizes in resident and migrant species breeding on farmland in Britain*

Clutch size	No. of species	
	Residents	Migrants
2	0	1
3	1	0
4	9	4
5	11	3
6	2	2
7	0	0
8	2	0
9	1	0
10	1	0
10+	4	1

Note: After pooling to avoid low expected frequencies for the migrants a χ^2 test on the relative frequencies of 4 or fewer and of 5 or more eggs per clutch gave $\chi^2 = 0.18$, n.s.

Fig. 9. Distributions of egg productivities (eggs per season) for resident (upper figure) and migrant (lower figure) species. The distributions do not differ significantly (Median test $\chi^2 = 0.49$, n.s.).

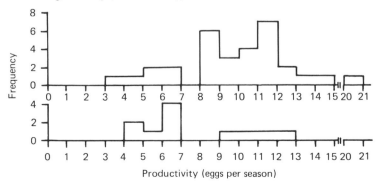

Table 7. *The distribution of breeding attempts each year in resident and migrant species*

| | No. of species | |
Clutches per year	Residents	Migrants
1	12	4
1+[a]	2	4
2	9	2
2+[b]	8	1
		$\chi^2 = 1.49$, n.s.[c]

[a] Some of the species population genuinely double-brooded.
[b] Some of the population genuinely treble-brooded.
[c] After pooling classes to avoid low expected frequencies (Siegel, 1956).

Fig. 10. Relationships between ubiquity (number of *Atlas* squares) and egg productivity for residents (upper figure) and for migrants (lower figure). For residents the two variables are uncorrelated ($r = 0.218$, n.s.) but for migrant species the regression line is significant: $y = 2681.1 + 238.4\,x$ with $r = 0.707$ ($P < 0.05$).

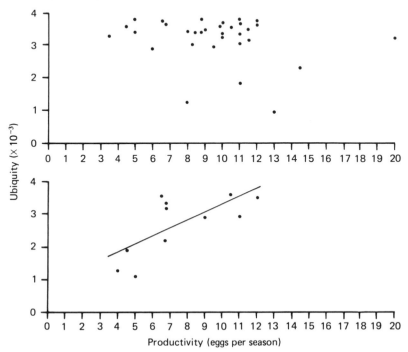

where Pr is egg productivity, whilst for migrant species the regression obtained was

$$\text{frequency} = 5.85 - 0.05\,Pr \qquad r = -0.110, \text{ n.s.}$$

showing that egg productivity had little effect on the likelihood of population increase or decrease occurring. Comparison of the two regressions showed no statistical differences between the two slopes $(F(1,35) = 0.74, \text{ n.s.})$. Examination of the effect of the size of population increases when they occurred in relation to egg productivity showed that the latter was also unimportant in determining the size of the population increase. For residents the regression was

$$\text{increase} = 11.25 - 0.11\,Pr \qquad r = -0.067, \text{ n.s.},$$

whilst for migrant species it was

$$\text{increase} = 14.41 + 0.56\,Pr \qquad r = 0.193, \text{ n.s.}$$

The size of population increases was thus independent of egg productivity in both groups.

Several theoretical studies have shown that there should be an inverse relationship between egg productivity and survival (Cole, 1954; Williams, 1966; Cody, 1971; Stearns, 1976) and several avian studies have shown that birds deplete their energy reserves in the course of producing more rather than fewer eggs (Schifferli, 1976; Jones & Ward, 1976). Fig. 11 shows that the same trade-off between survivorship and egg production was present amongst the migrant species $(r = -0.740, P < 0.01)$. Within the resident species the negative trend is only weakly present $(r = -0.214, \text{ n.s.})$. The difference in regression slope between the two groups is significant $(F(1,35) = 6.19, P < 0.05)$, showing that migrants suffer a greater reduction in achieving a given increase in egg productivity than do resident species.

Resilience

The concept of resilience is one of ability to recover from a population perturbation. In the present study it is possible to assess this quantitatively only for the resident species, by using the population depression brought about by the succession of two severe winters in 1961–2 and 1962–3 (Dobinson & Richards, 1964). Migrant species were not affected by this winter, of course, and lack a similarly uniform perturbation allowing comparative analysis.

For the resident species the resilience index was examined in relation to survival and to egg productivity. Regressing resilience on survival gave the equation

$$\text{resilience} = 46.65 - 27.66\,S \qquad r = -0.135, \text{ n.s.},$$

showing that the ability of these species to recover from the effects of hard winter was not determined by their survivorship.

Capacity for population recovery was, however, strongly correlated with species egg production (Fig. 12). That is, those resident species which had the greatest annual production of eggs were the species which recovered their original population most rapidly whilst those species with lower egg productivity recovered more slowly. Since survivorship and egg production were negatively correlated even amongst resident species (Fig. 11) partial correlations between resilience and survivorship and egg productivity with the other variable controlled were calculated. The effect of egg production on resilience remained even when survival was controlled ($r_{pr} = 0.664$, $P < 0.01$) whilst survival had no effect on resilience when egg productivity was controlled ($r_{pr} = 0.012$, n.s.). These results confirm that resilience was determined only by egg productivity and not at all by survivorship.

Fig. 11. Survival index in relation to egg production for residents ($r = -0.214$, n.s.) and for migrants ($r = -0.740$, $P < 0.05$), upper and lower graphs, respectively. For the latter the regression equation obtained is $y = 0.675$ $0.034 x$.

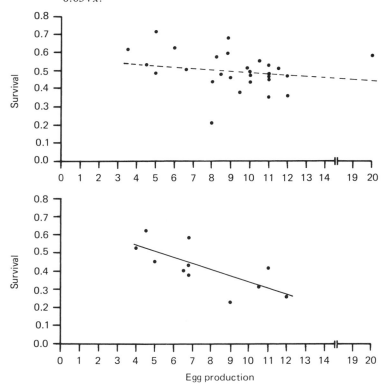

Discussion

The results above showed that migrant–resident differences occurred to a significant extent in three areas: (1) in body size and the correlations of survival with sexual dimorphism; (2) in the dependence of distribution and abundance on survival and egg productivity; and (3) in the variance of year-to-year population changes. The findings in these three areas can be drawn together within a single hypothesis, that resident species are strongly competitive with population levels regulated by climate in winter and by density-dependent behaviour in summer whilst migrant species are primarily exploiters of breeding season resources under-exploited by a resident population held down by winter mortality (Herrera, 1978).

Resident species were only slightly more sexually dimorphic in wing length than were migrant species (Table 5) but differed significantly from the latter in the dependence of survival rates on the extent of this dimorphism (Fig. 7). The difference is probably associated with the opposing selective pressures on dimorphism in the two groups. Several authors have suggested that sexual dimorphism allows males and females to exploit different parts of the spectrum of available resources, thus allowing the species a wider niche and greater total population levels than would be obtained if the two sexes shared a common niche (Selander, 1966). An alternative process suggested by Garnett (1976) would also result in selection for increased size dimorphism. Garnett suggests that in the great tit *Parus major* selection for large body size takes place in winter when body size confers a high dominance status and greater access to winter food

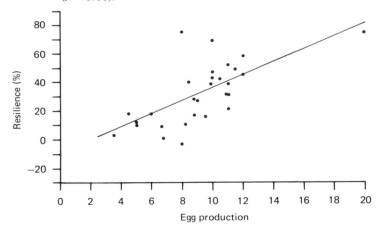

Fig. 12. Resilience (as percentage change in the farmland Common Birds Census index between 1963 and 1964) in relation to egg production, for residents species only (see text). The regression is $y = 4.50x - 9.03$ with $r = 0.672$ ($P < 0.01$).

through successful supplanting. In early spring, however, smaller females can breed earlier and, in this species, consequentially more successfully than can larger females. The greater reproductive success experienced by small females may offset the loss of fitness associated with small size in winter. The net result would be to select for large males and for small females, i.e. for increased sexual dimorphism. Garnett (1976) and Prys–Jones (1977) provide evidence for this model for great tit and reed bunting *Emberiza schoeniclus* respectively. Assuming only that resident species shared broadly similar ecological pressures, either model would generate a positive correlation between species survival and the extent of size dimorphism.

Migrant species may, in principle, experience similar pressures to those incorporated in the two models just discussed, the fact of their winter quarters lying outside the British Isles being largely irrelevant. But conditions during migration may constrain the extent to which wing dimorphism may evolve. In general, wing shape is correlated with migratory tendencies, with long-distance migrants possessing longer and more pointed wings (Ticehurst, 1938). It is reasonable to infer that this feature is correlated with efficiency of migratory flight. It follows, then, that in a sexually dimorphic migrant species either one sex has optimum wing length and the other an inferior one or both sexes diverge to a smaller extent, but in opposite directions, from the optimal length. Clearly, the greater the dimorphism the greater the average reduction in migratory efficiency in the species. To this extent the correlation of survival with dimorphism found here (Fig. 7) is explained. Moreau (1972) has already noted that for most palaearctic migrants reaching Africa, return northwards in spring involves long flights against head winds, conditions selecting for optimum aerodynamics of body shape and size.

Cox (1968) has considered how interspecific and intraspecific competition might influence the evolution of habitat and other resource use by a species in relation to its competitive ability, suggesting that competition might select for migration by particular species. If interspecific competition was not important in the evolution of migration, residents and migrants should be similar in the incidence of interspecific differentiation of feeding morphology. Cox found, however, that amongst various groups of North American birds the coefficient of variation of culmen length decreased the greater the proportion of migrant species in the group. Migration was thus most prevalent in those groups which were either unable, or were not compelled, to differentiate in morphology (Cox, 1968). This finding is relevant to the hypothesis above, that migrants were unable to exploit the strategy of sexual niche segregation because of the constraints of conditions during migration. This hypothesis might be tested by examining sex differences in survivorship in relation to dimorphism in wing length in migrants and in residents and perhaps by examining dimorphism within partial migrants in relation to the movements of the populations concerned.

Southwood (1976, 1977) has developed the concept of bionomic strategies (syntheses of such life-history features as size, longevity, fecundity, range and migration habit) and shown how the habitat of an animal can be regarded as a templet against which evolutionary forces fashion such ecological strategies. Of the five life-history features mentioned, migration habit is perhaps the one least studied to date, though the pioneering work of Taylor & Taylor (1977) indicates one approach to the interface between migration and population dynamics, and Southwood (1962) and others have previously considered for insects the relationship of the incidence of migration to habitat features.

Size figures prominently in the bionomic strategies considered by Southwood (Southwood *et al.,* 1974; Southwood, 1976) and in the present study size differences between migrants and residents were apparent (Fig. 6). Southwood argued increase in size as a major element within the 'feed-back loop' of so-called K-selection: larger size confers increased longevity and range, thus allowing the animal to integrate more of the fluctuations of its environment, so that the variance of its reproductive success is decreased; in addition, the increased longevity is accompanied by a corresponding decrease in fecundity and greater emphasis on maintaining adult survival (Cole, 1954), e.g. by increased predator avoidance. Species utilizing this type of strategy will normally exist in a stable habitat whose resources it is important not to overshoot. Consequently, emphasis within the strategy will be on maintaining a population at or just below the carrying capacity of the habitat and the increased survival of the birds living there is more adaptive than the recruitment of additional individuals to join the population in it. According to Southwood, in its simplest form the greater survival of the residents may simply derive from their knowledge of the geography of the habitat with its corresponding advantages for feeding and predator avoidance (Southwood, 1977). The present finding of a greater range of body size amongst British resident-bird species is indeed correlated with an emphasis on greater survival, as shown by the positive relationship between ubiquity (and thus abundance) and survival (Fig. 8).

The migrant species have features more akin to those of so-called 'r-selection'. Southwood (1977) presents a discussion geared primarily (though not exclusively) to the situation of insects or other short-lived animals born in one relatively ephemeral patch of habitat and forced to move to another patch of such habitat to breed successfully. This situation does not exactly parallel the position of migrant birds in Britain, involved as they are in a regular seasonal shift of population centres between the breeding season and the non-breeding season. Nevertheless, the northwards spring migration shares many of the requirements of Southwood's arguments: the birds will require ability to migrate, to survive the dangers of migration, and to arrive in the new habitats and breed there effectively. Whilst the detailed mechanisms of migration are beyond the scope of the present paper the breeding cycle does seem to be somewhat accelerated amongst migrants (J. J. M. Flegg & N. Riddiford, unpublished). In the present study, no net increase in egg productivity in migrants as against residents was detected (Fig. 9) but the ubiquity (and by inference the abundance) of migrants was found to be correlated with egg productivity (Fig. 10) and not with survival (Fig. 11). In many animals, increased emphasis on reproductive effort is frequently correlated with small body size (Southwood *et al.*, 1974), a point paralleling the small size of migrants noted in the present study.

These arguments suggest that resident birds in Britain are primarily K-selected whilst migrant species are r-selected. However, both groups studied here were breeding on farmland in Britain, so their differences in r–K selection cannot be associated with differences in habitat. It seems more likely that the difference is linked to seasonal variation in resource levels for breeding. MacArthur (1959) examined the proportion of migrant individuals present in breeding-season censuses in different parts of North America and suggested that the proportion of migrants was greatest wherever the change between winter and summer was greatest with respect to food supply. He noted that migrant individuals were most numerous in northern undisturbed habitats and failed to find any link between the proportion of migrants and the climate at the site. Willson (1976) has more recently reviewed MacArthur's analysis and offered some re-interpretation of his work. By examining the proportions of migrants of all types (thus extending the analysis beyond the study of purely tropical migrants to North America undertaken by MacArthur) she showed that habitat differences were less important than originally suggested by MacArthur. She suggested that latitudinal differences in seasonality were more probably a function of climate than of habitat type. MacArthur's and Willson's arguments thus suggest that migrant individuals would be more prevalent wherever there were large variations in insect food supply in the breeding season as a result of the prevailing climate. If true, this hypothesis would help explain the pattern of differences established in the present study and, in fact, evidence in support of climatic determination of winter–summer differences in resources availability for European migrant birds has recently been obtained (Herrera, 1978).

Herrera found a conspicuous gradient in the proportion of passerine breeding communities contributed by tropical migrants in Europe, with communities located in the northernmost areas showing the greatest percentage and those located to the south the smallest percentages. Whilst latitude was the major explanatory variable of this proportion, Herrera found that virtually the same result was obtained by considering the temperature of the coldest month in the year. Moreover, the trend was independent of habitat type but the structurally most diverse habitats (mixed coniferous–deciduous forests) tended to have fewer migrant individuals than did the structurally simpler habitats (fields, tundra, etc.). The latter point is in line with Southwood's (1977) argument that the greater 'adversity' in simple habitats provides more scope for r-selected species.

The findings of the present study provide support for a model developed by Herrera to account for the latitudinal/climatic gradient he found in the proportion of migrant species present. The model suggests that each breeding area has a particular carrying capacity as to the number of individual breeding birds it can support, irrespective of whether these are resident or migrant species. The number of resident individuals available to take up this carrying capacity should be determined by their overwintering survival: where survival is high many resident individuals survive to enter the breeding population the following spring; where survival is low, few residents survive to enter the breeding population. The model postulates that migrant individuals enter the breeding area to take up the unused resources available. In this way, breeding areas in a severe climate have relatively low overwintering survival of resident birds and provide considerable scope for migrants to breed, whilst in milder climates, residents overwinter more successfully and allow smaller incursions of migrant individuals. Herrera's models thus predicts that the abundance of resident birds should be correlated with between-area differences in survival. In the present study the abundance of resident individuals in a single area – that provided by British farmland – showed interspecific correlation between abundance and survival (Fig. 11), a result which might be anticipated on the basis of Herrera's argument. Conversely, one would not expect any relationship between abundance and survivorship in migrant species, and this was the case (Fig. 11). Again, Herrera predicts that most migrant individuals will be breeding in areas of maximum unused resources. The finding here (Fig. 10) that the most ubiquitous and abundant migrant species were those producing most eggs (and thus using most resources) is consistent with this argument. Finally, one might expect that if climatic factors are limiting in wintertime there should be greater scope for sexual divergence in feeding ecology, so as to exploit a greater range of the available food items, with concomitant increase in survivorship. This indeed was the case (Fig. 7)., though I note that the argument is weak because climatic limitation could operate via food availability.

The model suggested by Herrera makes few explicit predictions about the nature of population fluctuations in residents as against migrants. However, it is a fair assumption that if the resident species are indeed K-selected whilst simultaneously subject to a degree of regulation by winter climate, one would expect a rather more stable population level than prevails amongst migrants. This argument presupposes that climatic regulation operates frequently enough to prevent strong selection for density-dependent regulation: May (1976) has shown that excessively strong density dependence can lead to chaotic population fluctuations rather than to stability. Subject to this proviso, though, we would expect to find smaller increases and decreases of population amongst residents than amongst migrants, such as were indeed found (Figs. 2 and 3). Moreover, if the resident populations are generally increasing slightly following the effects of a very severe winter, such as occurred in 1962–3, the density of residents on British farmland will in general have increased over the period 1966–77, thus making it more and more difficult for migrant species to breed in numbers in the same habitat. Such an effect could account for the greater frequency of population decreases found for the migrants (Table 4).

Much of this case turns on the reality of K-selection in resident species. The BTO data for some species have been analysed from the point of view of density dependence and regulation of populations. For yellowhammers *Emberiza citrinella* clutch size on farmland decreases with increase in farmland breeding density, so that a greater proportion of the birds breed in other habitats, notably woodland scrub and other edge habitats (O'Connor, 1981a). The study showed the presence of differential density-dependence of clutch size in the two habitats, so that birds were distributing themselves between the two habitats in such a way as to maintain equal reproductive fitness, either by breeding in optimal but crowded farmland sites or in less crowded but suboptimal woodland sites. Similarly, great tits breeding on farmland form an overspill population from the preferred woodland sites, into which the birds move differentially as population density there alters (Krebs, 1971; O'Connor, 1980). These movements were accompanied by density dependence in clutch size and reproductive success, in survivorship, and in dispersal tendencies (Krebs, 1971; O'Connor, 1980). Finally, in another farmland species, the kestrel *Falco tinnunculus,* density-dependent factors operate to the point of introducing non-breeding in crowded populations, probably the epitome of K-selected responses (O'Connor, 1981b). Thus, the evidence is available for a variety of species resident on farmland in Britain to demonstrate the reality of K-selection processes, such as were postulated in explaining the differences between residents as a group and the migrants as a group. This is perhaps confirmed by the analysis of the resilience index (Fig. 12) presented here, showing as it does that most resident species displayed population increases correlated with their reproductive output rather than with adult survival in a year in which populations had been drastically reduced by density-dependent factors.

The present study suggests that resident species in Britain are primarily K-selected species adopting a strategy of individual survival within the country and showing high competitive ability to exploit resources during the breeding season. Migrant species can thus breed in Britain only to the extent that breeding season resources are under-utilized by the residents. According to Herrera's model the major element in determining the success of the residents will be their overwintering survival. Certain studies have already shown that the year-on-year changes in population levels on farmland are correlated with the severity of the winter: results for the great tit and for the yellowhammer have already been published (O'Connor, 1980, and 1981a) and unpublished data for a variety of other species are available. Examination of these data shows that the sensitivity of the annual population changes in these species is correlated with body weight, smaller birds being more sensitive; and unpublished BTO data for the changes in population during the most recent severe winter, that for 1978–9, show that interspecific differences in population changes across this one winter were indeed correlated with body weight, being smallest in the larger species. There thus seems to be every reason to accept the concept of r- and K-selection as underlying the resident–migrant differences established here and to view Herrera's model as providing a first level of explanation of the evolution of these differences.

This study arose from a wider-ranging examination of the usefulness of BTO data in testing current life-history theories. I am grateful to S. M. Taylor for much of the initial assembly of the data for this examination and to C. J. Mead for reviewing and up-dating the information available for analysis. Much of the cost of the Common Birds Census, and the author's salary, was provided by the BTO under a contract with the Nature Conservancy Council and this support is gratefully acknowledged.

References

Bailey, R. S. (1967). An index of bird population changes on farmland. *Bird Study*, **14**, 195–209.

Baker, R. R. (1978). *The Evolutionary Ecology of Animal Migration*. London: Hodder and Stoughton.

Calder, W. A. (1974). Consequences of body size for avian energetics. In *Avian Energetics*, ed. R. A. Paynter, pp. 86–144. Cambridge (Massachusetts): Nuttall Ornithological Club.

Campbell, B. C. & Ferguson-Lees, I. J. (1972). *A Field Guide to Birds' Nests*. London: Constable.

Cody, M. L. (1971). Ecological aspects of reproduction. In *Avian Biology*, vol. I. ed. D. S. Farner & J. R. King, pp. 462–512. New York: Academic Press.

Cole, L. C. (1954). The population consequences of life history phenomena. *Quarterly Review of Biology*, **29**, 103–37.

Cox, G. W. (1968). The role of competition in the evolution of migration. *Evolution*, **22**, 180–92.

Dobinson, H. M. & Richards, A. J. (1964). The effects of the severe winter of 1962/63 on birds in Britain. *British Birds*, **57**, 373–434.

Emlen, S. T. (1975). Migration: orientation and navigation. In *Avian Biology*, vol. **5,** ed. D. S. Farner & J. R. King, pp. 129–219. New York: Academic Press.

Garnett, M. C. (1976). *Some Aspects of Body Size in the Great Tit.* Oxford: D. Phil. thesis.

Haldane, J. B. S. (1955). The calculation of mortality rates from ringing data. In *Proceedings of the XI International Ornithological Congress,* ed. A. Portman & E. Sutter, pp. 454–8. Basle: Birkhauser Verlag.

Harrison, C. (1975). *A Field Guide to the Nests, Eggs and Nestlings of European Birds.* London: Collins.

Herrera, C. M. (1978). On the breeding distribution pattern of European migrant birds: MacArthur's theme reexamined. *Auk,* **95,** 496–509.

Jones, P. & Ward, P. (1976). The level of reserve protein as the proximate factor controlling the timing of breeding and clutch-size in the red-billed quela *Quelea quelea. Ibis,* **118,** 546–74.

Krebs, C. J. (1972). *Ecology: The Experimental Analysis of Distribution and Abundance.* New York: Harper and Row.

Krebs, J. R. (1971). Territory and breeding density in the great tit, *Parus major* L. *Ecology,* **52,** 2–22.

Lack, D. L. (1943–4). The problem of partial migration. *British Birds,* **37,** 122–30, 143–50.

MacArthur, R. H. (1959). On the breeding distribution pattern of North American migrant birds. *Auk,* **76,** 318–25.

Matthews, G. V. T. (1955). *Bird Navigation.* Cambridge: Cambridge University Press.

May, R. M. (1976). Models for single populations. In *Theoretical Ecology,* ed. R. M. May, pp. 4–25. Oxford: Blackwell Scientific Publications.

Moreau, R. E. (1972). *The Palearctic–African Bird Migration Systems.* London: Academic Press.

O'Connor, R. J. (1980). Pattern and process in great tit *(Parus major)* populations in Britain. *Ardea,* **68,** 165–83.

O'Connor, R. J. (1981*a*). Population regulation in the yellowhammer *Emberiza citrinella* in Britain. In *Proceedings of the VI International Bird Census Committee Conference,* ed. H. Oelke. Gottingen: IBCC.

O'Connor, R. J. (1981*b*). The breeding dynamics of British kestrels. *Ibis,* **122,** 420–1.

Prys-Jones, R. P. (1977). *Aspects of Reed Bunting Ecology, with Comparisons with the Yellowhammer.* Oxford: D. Phil. thesis.

Schifferli, L. (1976). *Factors Affecting Weight and Condition in the House Sparrow, Particularly when Breeding.* Oxford: D. Phil. thesis.

Selander, R. K. (1966). Sexual dimorphism and differential niche utilization in birds. *Condor,* **68,** 113–51.

Sharrock, J. T. R. (1976). *The Atlas of Breeding Birds in Britain and Ireland.* Tring: British Trust for Ornithology.

Siegel, S. M. (1956). *Nonparametric Statistics for the Behavioral Sciences.* New York: McGraw–Hill.

Southwood, T. R. E. (1962). Migration of terrestrial arthropods in relation to habitat. *Biological Review,* **37,** 171–214.

Southwood, T. R. E. (1976). Bionomic strategies and population parameters. In *Theoretical Ecology,* ed. R. M. May, pp. 26–48. Oxford: Blackwell Scientific Publications.

Southwood, T. R. E. (1977). Habitat, the templet for ecological strategies? Presidential address to the British Ecological Society, 5 January 1977. *Journal of Animal Ecology,* **46,** 337–65.

Southwood, T. R. E., May, R. M., Hassell, M. P. & Conway, G. R. (1974). Ecological strategies and population parameters. *American Naturalist,* **108,** 791–804.

Spencer, R. & Hudson, R. (1978). Report on bird-ringing for 1976. *Ringing & Migration,* **1,** 189–252.

Stearns, S. C. (1976). Life-history tactics: a review of the ideas. *Quarterly Review of Biology,* **51,** 3–47.

Taylor, L. R. & Taylor, R. A. J. (1977). Aggregation, migration and population mechanics. *Nature, London,* **265,** 415–21.

Taylor, S. M. (1965). The Common Birds Census – some statistical aspects. *Bird Study,* **12,** 268–86.

Ticehurst, C. B. (1938). *A Systematic Review of the Genus* Phylloscopus. London: British Museum (Natural History).

Williams, G. C. (1966). Natural selection, the costs of reproduction and a refinement of Lack's principle. *American Naturalist,* **100,** 687–90.

Williamson, K. & Homes, R. C. (1964). Methods and preliminary results of the common birds census, 1962–63. *Bird Study,* **11,** 240–56.

Willson, M. F. (1976). The breeding distribution of North American migrant birds: a critique of MacArthur (1959). *Wilson Bulletin,* **88,** 582–87.

Winstanley, D., Spencer, R. & Williamson, K. (1974). Where have all the whitethroats gone? *Bird Study,* **21,** 1–14.

Witherby, H. F., Jourdain, F. C. R., Ticehurst, N. F. & Tucker, B. W. (1938). *The Handbook of British Birds.* London: Witherby.

T.R.E.SOUTHWOOD

Ecological aspects of insect migration

Migration is one of those terms whose general meaning is widely understood, but whose precise definition is difficult. In this Symposium we have considered migrations of organisms ranging in size from aphids to whales and in social complexity from butterflies to man and this has revealed the diversity in the functional expressions of migration. The major attempt to bring these together within one framework is that of Baker (1978). It is no criticism of Baker, but a reflection of the nature of the phenomenon, that his definition of migration, 'the act of moving from one spatial unit to another', is so broad as to be almost indistinguishable functionally from movement of the whole animal.

In his *magnum opus* on insect migration, Johnson (1969) highlighted two criteria for the recognition of migration, one was behavioural – undistracted movement and the other ecological – departure from the habitat. The recognition of habitat again presents problems, fully explored by Baker (1978), but the definition 'the area that provides the resource requirements for a discrete phase of an animal's life', encapsulates what most of us understand by the term. The habitat, as so defined, is also the area traversed by the animal's trivial movements, that is those within its sensory range where it forages (Hassell & Southwood, 1978).

Another criterion is the degree of scattering, the increase in the mean distance ($\bar{r}$) between individuals: this process is properly termed dispersal. Dispersal and migration are in insects often, but by no means always, synonomous. Their relationship is shown in Fig. 1.

It will be obvious that there are problems in the clear-cut definition of migration and these are elaborated by Baker (1978): the logical elimination of these difficulties leads to his definition which, as I have said, I regard as unhelpfully broad. He concludes his analysis with an important point: the typical question he writes is 'Why does animal A migrate?' and indeed the provisional title I was allotted was, 'Why insects migrate'. Baker however points out that the real question is 'Why does animal A migrate and not animal B?' Specifying insect as the animal this could be taken as the subtitle of this article.

Expression of migratory ability

A comparison of the level of migration, both between species and between individuals of the same species, is aided if certain traits, that may be measured, can be associated with migration. Much evidence is based on such indirect evaluations and so these traits will be briefly enumerated.

In insects most, but by no means all, migration is by flight and thus the absence, or reduction to a non-functional size, of wings is generally taken as an indication of a non-migrant. Shaw (1970*a,b,c*) showed in *Aphis fabae* that individual differences in actual wing size are related to 'migratory urge'; the apterous individuals represent the extreme non-migrant condition. However the presence of full-sized wings must not be taken as evidence of flight ability, let alone migration, for wing muscles may never develop (e.g. in many aquatic Coleoptera (Jackson, 1956)) or be lost during life (e.g. in aphids (B. Johnson, 1953), bark beetles (Chapman, 1956) and corixid water bugs (Young, 1965)).

Migration demands sustained locomotor activity (Kennedy, 1951, 1961) and the migratory potential of structurally comparable insects can sometimes be discerned from their reserves: a large amount of stored fat is often characteristic of a migrant, and more particularly one that enters diapause after migration (Fedotov, 1947; Atkins, 1969).

In evaluating flight as an indicator of migration it is useful to distinguish between those insects like most Diptera in which it serves for both trivial and

Fig. 1. The terms 'migration' and 'dispersal'.

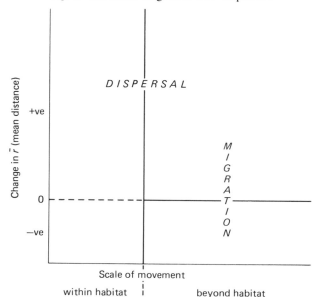

migratory movements, and those like most Coleoptera in temperate regions, where walking is the normal means of locomotion for trivial movement, and flight serves for migration (Southwood & Johnson, 1957). In aquatic Coleoptera and Hemiptera this distinction is complete; but in tropical and subtropical regions, whether as a result of selection by predators or purely a physiological response to temperature is unclear, flight serves widely for trivial movement in most taxa.

Ballooning on silk threads and phoresy are other methods of migration in terrestrial arthropods, whilst the first instar larva of the salt marsh aphid *Pemphigus trehernei*, has been shown to migrate by floating on the sea surface (Foster & Treherne, 1978).

Let us now consider the actual movement. As Kennedy (1961, 1975) has pointed out migration has certain characteristic features, not only persistent locomotor activity, but also straightening-out movements and the depression of vegetative responses with particular take-off and alighting patterns. But consideration of the behavioural aspects has led to controversy as to the extent to which the migrating insect is wind-born like an inert particle or maintains a flight direction. The living insect is never analogous to the inert particle; even with a non-flying aphid the terminal velocity can be influenced by the angle at which the wings are held (even though they are not moved (Thomas, Ludlow & Kennedy, 1977)). Thus I believe that even the most feeble flyer should be regarded as more resembling a canoeist travelling downstream than a piece of flotsam (Southwood, 1978a): whilst for stronger flyers cross-current or even up-current navigation is possible.

There is now definite evidence of orientation with larger, more strongly flying insects, that may keep close to the ground (Baker, 1968a, b; Sotthibandhu & Baker, 1979): thus direction may be virtually independent of wind. These observations mostly on Lepidoptera, should not, I believe, be regarded as placing them in a different category from other insects. The difference is one of degree rather than kind. Given the relative flight strengths of different groups and the average wind speeds close to the ground it is clear that, whereas directional migration is an evolutionary option for many Lepidoptera, minute insects, for example winged aphids and cecidomiid midges (Sylven, 1970), are constrained to be 'down-stream canoeists' for most of the time and evolutionary forces would operate to perfect these tactics for the respective groups. Furthermore as Baker (1968a, 1978) has pointed out orientation, irrespective of direction, increases the efficiency of search for habitats and, in most species, it is not time compensated so that corrections are not made for celestial movements. Thus a migration will not terminate at a point towards which a non-time compensated celestially orientating butterfly was heading at the outset. If there is some further displacement, greater or lesser, due to wind direction this is not significant from either the ecological or evolutionary viewpoints: the canoeist analogy still holds. Thus beside the evidence of Baker (1978) and colleagues on Lepidoptera orientation, we may place, without conflict so far as their ecological function is concerned, the observations of French (1963, 1969) and Mikkola (1967, 1970, 1971) that long-distance movements of other Lepidoptera are wind-born. This is not to imply that suitable habitats are randomly distributed as to compass direction. For many temperate species they lie to the north and/or west in the spring; Baker (1968a, 1978) has evidence of change in compass direction orientation between spring and autumn; winds and other weather features also tend to have patterns whose nature and probability will be allowed for by selection and as Schaefer (1976) has shown, take-off may be frequently influenced by wind direction. Johnson (1969) reviews the close relationship between weather patterns and the movement of insects north and south through the central region of the USA. The phenomenon of the carriage of locust swarms on winds to zones of convergence and hence of rainfall demonstrated by Rainey (1951) is ecologically identical, although the relation to compass direction is variable. The rice leafhoppers (*Sogatella* and *Nilaparvata*) appear to migrate annually from central China to Japan utilizing depressions that move from the southwest: variations in pest level seem to depend on yearly fluctuations in these weather patterns (Kisimoto, 1976).

The migration incidence or innate level of the species

An understanding of the evolutionary significance of any behavioural or physiological mechanism is facilitated if its expression in different species varies, and may be related to their ecology.

There is now a wide body of evidence that relates the incidence of migration in a species to the nature of its habitat (Southwood, 1962; Sweet, 1964; Johnson, 1969; den Boer, 1970; Roff, 1974, 1975; Baker, 1978; Denno & Grissell, 1979). The incidence is highest in those species whose habitats are temporary, relatively short-lived in a particular location; i.e. of a low durational stability (Southwood, 1977). As Baker (1978) has argued, correctly I believe, from the evolutionary viewpoint more emphasis should be placed on the appearance of new habitats (and on their high suitability when they first arise) than on their longevity: the real selective advantage is to colonize a new habitat when it is an 'ecological vacuum'; at this time the intrinsic rate of natural increase will have a high value.

Insects are frequently polymorphic in respect of migrating incidence; this most frequently manifests itself structurally in wing and body form (Richards, 1961; Young, 1965a, b; Acton & Scudder, 1969; Vepsäläinen, 1971, 1978; Andersen, 1973; Waloff, 1973; Y. Y. May, 1975; Jarvinen & Vepsäläinen, 1976; Iheagwam, 1977; Dingle, 1978; Denno, 1976; Denno & Grissell, 1979) and the whole topic has currently been comprehensively reviewed by Harrison (1980). The frequency of morphs may be determined solely by a genetic mechanism, but more generally this is modified by environmental conditions – season, weather, food supply, crowding. The resulting mix of morphs may be considered to represent the 'best' in relation to the features of the species habitat – a 'best' arrived at through selection (Southwood, 1977). In the studies referred to above it was generally possible to relate migratory ability to habitats of low durational stability; that is to situations where there is a high probability of finding a new habitat. As Southwood (1977) and Baker (1978) have both stressed, evolutionary selection balances the risks of migration (including the probability of finding a new habitat) against the risks of remaining in the old.

Immediate factors influencing the level of migration

An early view of migration was that it was a response to a major disaster and, as with the lemming, appeared to end in catastrophe. In 1960 C. G. Johnson wrote 'I see migration as an evolved adaptation rather than as a current reaction to adversity' and thereby heralded an evolutionary approach to its significance. The incidence of migration was related to the characteristics of the habitat that had fashioned its evolution, as outlined above, and the significance of immediate events appeared to be denied. It was more a matter, as so often in the history of development of scientific theories, of concentration on a new aspect and neglect of other facets, than a conviction that immediate factors were immaterial (e.g. Southwood (1962, p. 201) wrote 'the correlation between overcrowding and migration, sometimes true, sometimes false . . .'). However, others were soon to point out many instances where current adversity was related to migration (e.g. Dixon, Burns & Wangboonkong, 1968). Indeed there is really no conflict between the concept that the mechanisms required for migration are evolved adaptations and that the immediate (current) factors will trigger their operation. The extent to which immediate factors are involved will also be influenced by evolution: if there is little variability in the ontogeny of habitat suitability (Baker, 1978) then migration will be obligatory for those individuals capable of migration (their proportion may be high or low). If there is wide variation in habitat suitability (at the same ontogenetic stage) then migration will be facultative. Thus again a controversy reveals itself to be merely considering the two ends of a spectrum. The cues may operate at various times before the actual migration; crowding, food quality, weather and season commonly serve as cues for the facultative initiation of migration (Johnson, 1969; Baker, 1978).

The role of migration in population processes

Migration, as we have seen, has evolved as an adaptation to environmental heterogeneity in space and time; its onset may be signalled by immediate factors. I will now consider how this process impinges on populations. The effect of migration is that animal populations are continually redistributed; the greater the level of migratory activity (the incidence of migration) in the species the more rapidly its distribution pattern will change. Taylor & Taylor (1977, 1978, 1979) have shown how the distribution of Lepidoptera in Britain, as monitored by light traps, changes from season to season.

The scale of this redistribution varies greatly, being related to the features of the insects' habitat. An excellent example of the small-scale pattern change is provided by Whittaker, Ellistone & Patrick's (1979) study on the beetle, *Gastrophysa viridula* on dock plants around a river; both on the river banks and on a shingle island. On *four* occasions over a mere *three* years the population on the island was eliminated by floods: on only one (further) occasion did some eggs survive flooding. During the summer months the island was recolonized within weeks.

At the other extreme there are large-scale movements: these are well known in the monarch butterfly (*Danaus plexippus*) in North America and in locusts in Africa. The Australian bushfly (*Musca vetustissima*) has now been demonstrated to move over a large range (Hughes & Nicholas, 1974), re-invading southeastern Australia every spring in a series of successive waves.

In these species, and in the hundreds of others that could be given as examples, colonization, or rather recolonization, is a regular part of the life-time track (as Baker (1978) terms it) of the individual. Evolution would be expected to operate to give the migrating individuals a high reproductive value (Dingle, 1965) and indeed migration is regarded as a part of the 'r-strategy' (Dingle, 1972, 1974; Pianka, 1970; Southwood, 1975). However as Kennedy (1975) has pointed out, a paradox is involved because the act of migration defers reproduction – the feature of the r-strategy: but this deferment of migration is a physiological necessity for them to reach the new habitat wherein their reproductive rate made by adopting the migration tactic is, on the average, more than compensated for by the enhanced survival provided by the new habitat. In most insects the value of r for the migrant is reduced not only by the time taken to migrate, but also because in polymorphic species the actual fecundity is lowered (Johnson, 1969; Dixon, 1972; Baker, 1978). A striking exception is provided by V. K. Taylor's (1978) observations on the featherwing beetles, *Ptinella,* the alatae of which have a higher oviposition rate than the apterae. One can only suppose that the apterae have some advantage, perhaps in seeking out food resources, over alatae in the competitive environment of a colonized log: an advantage linked in some way or another with reduced fecundity.

The migrants' reproductive potential as colonists is commonly enhanced by various adaptations that increase the probability of females being fertilized. In many species, those conventionally regarded as migrants, such as the monarch butterfly or locusts, males and females migrate together. In others the females may commonly be fertilized before migration, a process where probability may be enhanced by the males remaining around the emergence site in the original habitat. Adiseyun & Southwood (1979) have shown this to apply with the frit fly (*Oscinella frit*) which migrates from grassland to oat crops in the spring (Fig. 2). In the featherwing beetle, *Ptinella aptera* the volume of the spermatheca of the alate is nearly twice the size of that in the apterous individuals (V. K. Taylor, 1978), thereby ensuring that the fecundity of the alate female can be realized in a male-less environment.

Role of migration in zoogeography

Besides its major ecological role, the discovery of newly developed habitats within the species range (e.g. the fresh dung pat, the new pool or the fresh cabbage field), discussed above, migration is also an important contributor to zoogeographical patterns. On a geographical scale invasions will range from the colonization of off-shore islands (Lindroth, 1971; Simberloff, 1978) to the crossing of major oceans (Johnson & Bowden, 1973). Major displacements are often associated with storms (Tulloch, 1929; Mitchell, 1962), but the insects involved are usually those with particularly high incidences of migration (Southwood, 1962; Howden, 1977; Cheng & Birch, 1978).

Conclusion

The ecological significance of migration resides in enabling animals to have lifetime tracks that run through a series of habitats, each one of which is suitable for only part of its life. With the development of wings, insect species have been well able to exploit this strategy and I believe it is one of the factors that contributes to the large number of species of insects (Southwood, 1978*b*). The physiological and behavioural adaptations associated with the onset of migration may be seen to represent sensitive mechanisms for responding to the decay of one habitat and, once migrating, maximizing the probability of discovering a new one. The considerable variations found in these mechanisms must be regarded as the evolutionary response to variations in habitat development patterns, in the scale of the movement and in the size of the insect.

These variations are especially pronounced in the extent to which the animal provides, from its own muscles, the power for the movement or selectively utilizes for this purpose the ambient medium. The selective tidal transport exhibited by flat fish (Arnold, this volume) is exactly analogous to the use of air currents (under certain weather conditions) by aphids, locusts and other insects. The most powerful animals, whales, pigeons and salmon are examples discussed in this Symposium, will rely on their own locomotion and hence will evolve sophisticated mechanisms for orientation. The relative importance of the two will vary (Fig. 2). As in most evolutionary situations there will be a spectrum; it is probable, I believe that the evolutionary advantage will lie towards the extremes and that if species numbers were plotted against the horizontal (x) axis the distribution would be found to be bimodal.

However the optimal evolved strategy will not be determined simply by the relative power of the animal and the movement of the ambient medium. The extent to which the new habitats lie in one direction and the extent to which the ambient medium moves directionally will be important influences and may be termed 'directionality values'. The more precisely directional the sites of the new habitats the stronger the evolutionary forces for the development of an ability to orientate, whilst the more directional the movement of the ambient media the less evolutionary forces will act on the animal to promote the development of orientating abilities.

Fig. 2. The roles of movement and transport by air or water.

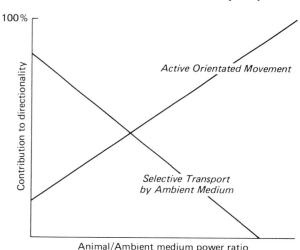

References

Acton, A. B. & Scudder, C. G. E. (1969). The ultrastructure of the flight muscle polymorphism in *Cenocorixa bifida* (Hung.) (Heteroptera: Corixidae). *Zeitschrift für Morphologie und Ökologie der Tiere*, **65**, 327–35.

Adiseyun, A. A. & Southwood, T. R. E. (1979). Differential migration of the sexes in *Oscinella frit* (Diptera:Chloropidae). *Entomologia Experimentalis et Applicata*, **25**, 59–63.

Andersen, N. M. (1973). Seasonal polymorphism and developmental changes in organs of flight and reproduction in bivoltine pondskaters (Hem: Gerridae). *Entomologica Scandinavica*, **4**, 1–20.

Atkins, M. D. (1969). Lipid loss with flight in the Douglas-fir beetle. *Canadian Entomologist*, **101** (2), 164–5.

Baker, R. R. (1968*a*). A possible method of evolution of the migratory habit in butterflies. *Philosophical Transactions of the Royal Society*, B. **253**, 309–41.

Baker, R. R. (1968*b*). Sun orientation during migration in some British butterflies. *Proceedings of the Royal Entomological Society of London*, (A) **143**; 89–95.

Baker, R. R. (1978). *The Evolutionary Ecology of Animal Migration*, London: Hodder & Stoughton.

Boer, P. J. den (1970). On the significance of dispersal power for populations of carabid beetles (Coleoptera:Carabidae). *Oecologia, Berlin* **4**, 1–28.

Chapman, J. A. (1956). Flight muscle changes during adult life in a scolytid beetle. *Nature, London*, **177**, 1183.

Cheng, L. & Birch, M. C. (1978). Insect flotsam: an unstudied marine resource. *Ecological Entomology*, **3**, 87–97.

Denno, R. F. (1976). Ecological significance of wing polymorphisms in Fulgoroidea which inhabit tidal salt marshes. *Ecological Entomology*, **1**, 257–66.

Denno, R. F. & Grissell, E. E. (1979). The adaptiveness of wing dimorphism in the salt marsh-inhabiting planthopper *Prokelisia marginata* (Homoptera: Delphacidae). *Ecology*, **60** (1), 221–36.

Dingle, H. (1965). The relation between age and flight activity in the milkweed bug, *Oncopeltus*. *Journal of Experimental Biology*, **42**, 269–83.

Dingle, H. (1972). Migration strategies of insects. *Science*, **175**, 1327–35.

Dingle, H. (1974). Diapause in a migrant insect, the milkweed bug *Oncopeltus fasciatus* (Dallas) (Hemiptera:Lygaeidae). *Oecologia, Berlin*, **17**, 1–10.

Dingle, H. (1978). Migration and diapause in tropical, temperate and island milkweed bugs. In *Proceedings in Life Sciences: Evolution of Insect Migration and Diapause*, pp. 254–76. New York: Springer-Verlag.

Dixon, A. F. G. (1972). Control and significance of the seasonal development of colour forms in the sycamore aphid, *Drepanosiphum platanoides* (Schr.) *Journal of Animal Ecology*, **41**, 689–97.

Dixon, A. F. G., Burns, M. D. & Wangboonkong, S. (1968). Migration in aphids: response to current adversity. *Nature, London*, **220**, 1337–8.

Fedotov, D. M. (1947). The noxious little tortoise, *Eurygaster integriceps* Put. Reports on the work of the expedition to Central Asia for the study of the noxious little tortoise organised by the A. N. Severtgov Institute of Evolutionary Morphology, vol. **1, 2**. *Moscow Akademia Nauk USSR., R.A.E.* **40**, 307.

Foster, W. A. & Treherne, J. E. (1978). Dispersal mechanisms in an intertidal aphid. *Journal of Animal Ecology*, **47**, 205–17.

French, R. A. (1963). Migration records, 1961. *Entomologist,* **96,** 32–8.

French, R. A. (1969). Migration of *Laphygma exigua* Hubner (Lepidoptera: Noctuidae) to the British Isles in relation to large-scale weather systems. *Journal of Animal Ecology,* **38,** 199–210.

Harrison, R. G. (1980). Dispersal polymorphisms in insects. *Annual Reviews of Ecology and Systematics,* **11,** 95–118.

Hassell, M. P. & Southwood, T. R. E. (1978). Foraging strategies in insects. *Annual Reviews of Ecology and Systematics,* **9,** 75–98.

Howden, H. F. (1977). Beetles, beach drift and island biogeography. *Biotropica,* **9,** 53–7.

Hughes, R. D. & Nicholas, W. L. (1974). The spring migration of the bushfly (*Musca vetustissima* Walk.): Evidence of displacement provided by natural population markers including parasitism. *Journal of Animal Ecology,* **43** (2), 411–28.

Iheagwam, E. U. (1977). Comparative flight performance of the seasonal morphs of the cabbage whitefly, *Aleyrodes brassicae* (Walk.) in the laboratory. *Ecological Entomology,* **2,** 267–71.

Jackson, D. J. (1956). The capacity for flight of certain water beetles and its bearing on their origin in the Western Scottish Isles. *Proceedings of the Linnean Society of London,* Session **167,** 1954–5, pt. 1, pp. 76–96.

Järvinen, O. & Vepsäläinen, K. (1976). Wing dimorphism as an adaptive strategy in water-striders (*Gerris*). *Hereditas* **84,** 61–8.

Johnson, B. (1953). Flight-muscle autolysis and reproduction in aphids. *Nature, London,* **172,** 813.

Johnson, C. G. (1960). A basis for a general system of insect migration and dispersal by flight. *Nature, London,* **186** (4722), 348–50.

Johnson, C. G. (1969). *Migration and Dispersal of Insects by Flight.* London: Methuen & Co.

Johnson, C. G. & Bowden, J. (1973). Problems related to the transoceanic transport of insects, especially between the Amazon and Congo areas. In *Tropical Forest Ecosystems in Africa and South America: A Comparative Review,* ed. B. J. Reggers, E. S. Ayensu & W. D. Duckworth, pp. 207–22. Washington: Smithsonian Institute.

Kennedy, J. S. (1951). The migration of the desert locust (*Schistocerca gregaria* Forsk.). I. The behaviour of swarms. II. A theory of long-range migrations. *Philosophical Transactions of the Royal Society,* B (B) **235,** 163–290.

Kennedy, J. S. (1961). A turning point in the study of insect migration. *Nature, London,* **189** (4767), 785–91.

Kennedy, J. S. (1975). Insect dispersal. In *Insects, Science and Society,* ed. D. Pimentel, pp. 103–19. New York: Academic Press.

Kisimoto, R. (1976). Synoptic weather conditions inducing long-distance immigration of planthoppers, *Sogatella furcifera* Hörvath and *Nilaparvata lugens* Stol. *Ecological Entomology,* **1,** 95–109.

Lindroth, C. H. (1971). Biological investigations on the new volcanic island Surtsey, Iceland. In *Dispersal and Dispersal Power of Carabid Beetles (Symposium of the Biological Station, Wijster,* Nov. 1969): Miscellaneous papers *Landbouw der Hogeschule der Wageningen,* **8,** 65–9.

May, Y. Y. (1975). Study of two forms of the adult *Stenocranus minutus. Transactions of the Royal Entomological Society of London.,* **127,** 241–54.

Mikkola, K. (1967). Immigrations of Lepidoptera, recorded in Finland in the years 1946–1966, in relation to air currents. *Annales Entomologic Fennici,* **33,** 65–99.

Mikkola, K. (1970). The interpretation of long-range migrations of *Spodoptera exigua* Hb. (Lepidoptera:Notuidae) *Journal of Animal Ecology*, **39**, 593–8.

Mikkola, K. (1971). The migratory habit of *Lymantria dispar* (Lep. Lymantriidae) adults of continental Eurasia in the light of a flight to Finland. *Acta Entomologica Fennica*, **28**, 107–20.

Mitchell, R. (1962). Storm-induced dispersal in the damselfly *Ischnura verticalis* (Say). *American Midland Naturalist*, **68** (1), 199–202.

Pianka, E. R. (1970). On r- and K- selection. *American Naturalist*, **104**, 592–7.

Rainey, R. C. (1951). Weather and the movement of locust swarms: a new hypothesis. *Nature, London*, **168**, 1057–60.

Richards, O. W. (1961). An introduction to the study of polymorphism in insects. *Symposia of the Royal Entomological Society of London*, **1**, 13–21.

Roff, D. A. (1974). Spatial heterogeneity and the persistence of populations. *Oecologia, Berlin*, **15**, 245–58.

Roff, D. A. (1975). Population stability and the evolution of dispersal in a heterogeneous environment. *Oecologia, Berlin*, **19**, 217–37.

Schaefer, G. W. (1976). Radar observations of insect flight. In *Insect Flight*, ed. R. C. Rainey, *Symposia of the Royal Entomological Society of London*, **7**, 157–97.

Shaw, M. J. P. (1970*a*). Effects of population density an alienicolae of *Aphis fabae*, Scop. 1. The effect of crowding on the production of alatae in the laboratory. *Annals of Applied Biology*, **65**, 191–6.

Shaw, M. J. P. (1970*b*). Effects of population density on alienicolae of *Aphis fabae* Scop. 2. The effect of crowding on the expression of migratory urge among alatae in the laboratory. *Annals of Applied Biology*, **65**, 197–203.

Shaw, M. J. P. (1970*c*). Effects of population density on alienicolae of *Aphis fabae* Scop. 3. The effect of isolation on the development of form and behaviour of alatae in a laboratory clone. *Annals of Applied Biology*, **65**, 205–21.

Simberloff, D. S. (1978). Colonisation of islands by insects: immigration, extinction and diversity. In *Diversity of Insect Faunas*, ed. L. A. Mound & N. Waloff, *Symposia of the Royal Entomological Society of London*, **9**, 139–53.

Sottibandhu, S. & Baker, R. R. (1979). Celestial orientation by the large yellow underwing moth. *Noctua pronuba* L. *Animal Behaviour*, **27**, 786–800.

Southwood, T. R. E. (1962). Migration of terrestrial arthropods in relation to habitat. *Biological Reviews*, **37**, 171–214.

Southwood, T. R. E. (1975). The dynamics of insect populations. In *Insects, Science and Society*, ed. D. Pimentel, pp. 151–99. New York: Academic Press.

Southwood, T. R. E. (1977). Habitat, the templet for ecological strategies? *Journal of Animal Ecology*, **46**, 337–65.

Southwood, T. R. E. (1978*a*). Escape in space and time – concluding remarks. In *Evolution of Insect Migration and Diapause*, ed. H. Dingle, pp. 277–9. New York: Springer-Verlag.

Southwood, T. R. E. (1978*b*). The components of diversity. In *Diversity of Insect Faunas*, ed. L. A. Mound & N. Waloff, *Symposia of the Entomological Society*, **9**, 19–40.

Southwood, T. R. E. & Johnson, C. G. (1957). Some records of insect flight activity in May 1954 with particular reference to the massed flights of

Coleoptera and Heteroptera from concealing habitats. *Entomologist's Monthly Magazine*, **93**, 121–6.

Sweet, M. H. (1964). The biology and ecology of the Rhyparochrominae of New England (Heteroptera, Lygaeidae). Part I. *Entomologica Americana* **43**, 1–124.

Sylven, E. (1970). Field movement of radioactively labelled adults of *Dasyneura brassicae* Winn. (Dipt., Cecidomyiidae). *Entomologica Scandinavica*, **1**, 161–87.

Taylor, L. R. & Taylor, R. A. J. (1977). Aggregation, migration and population mechanics. *Nature, London*, **265** (5593), 415–21.

Taylor, L. R. & Taylor, R. A. J. (1978). The dynamics of spatial behaviour. In *Population Control by Social Behaviour, Symposium of the Institute of Biology*, 181–212.

Taylor, R. A. J. & Taylor, L. R. (1979). A behavioural model for the evolution of spatial dynamics. In *Population Dynamics*, ed. R. M. Anderson & B. D. Turner, *Symposia of the British Ecological Society*, **20**, 1–27.

Taylor, V. K. (1978). A winged elite in a subcortical beetle as a model for a prototermite. *Nature, London*, **275** (5683), 73–5.

Thomas, A. A. G., Ludlow, A. R. & Kennedy, J. S. (1977). Sinking speeds of falling and flying *Aphis fabae* Scopoli. *Ecological Entomology*, **2** (4), 315–26.

Tulloch, J. B. G. (1929). Dragonfly migration. *Entomologist*, **62**, 213.

Vepsäläinen, K. (1971). The role of gradually changing day length in determination of wing length, alary dimorphism and diapause in a *Gerris odontogaster* (Zett.) population (*Gerridae:Heteroptera*) in South Finland. *Annales Academiae Scientiarum Fennicae*, (A), IV *Biologica*, **183**, 1–25.

Vepsäläinen, K. (1978). Wing dimorphism and diapause in *Gerris:* determination and adaptive significance. In *Evolution of Insect Migration and Diapause*, ed. H. Dingle, pp. 218–53. New York: Springer-Verlag.

Waloff, N. (1973). Dispersal by flight of leafhoppers (Auchenorrhyncha: Homoptera). *Journal of Applied Ecology*, **10**, 705–30.

Whittaker, J. B., Ellistone, J. & Patrick, C. K. (1979). The dynamics of a chrysomelid beetle, *Gastrophysa viridula*, in a hazardous natural habitat. *Journal of Animal Ecology*, **48**, 973–86.

Young, E. C. (1965a). Flight muscle polymorphism in British Corixidae: ecological observations. *Journal of Animal Ecology*, **34**, 353–96.

Young, E. C. (1965b). The incidence of flight polymorphism in British Corixidae and description of the morphs. *Journal of Zoology*, **146**, 567–76.

R.J.V.JOYCE

The control of migrant pests

Introduction

It is characteristic of all living organisms to disperse in order that they may colonize new or unexploited habitats, or to escape from a habitat no longer suitable for survival. The strategies of dispersal are legion, varying from the aerial transport of bacterial and fungal spores to the migration of birds. The latter, employing sophisticated navigation systems, are able to colonize their new habitat and survive as a species with a very low reproductive rate, whilst the success of the former depends on producing millions of progeny. Rainey (1973) provides examples, the wheat rusts *Puccinia graminis* and *P. triticina* in North America, the spores of which were found to be deposited 1 000 km from their source at a rate of 100 per sq. ft per 24 h, and the ruby-throated hummingbird *Archilocus colubris* L., which migrates annually over the same route laying a clutch of two eggs only annually.

This paper, in considering the control of migrant pests, will be concerned mainly with insects, which have evolved many strategies for dispersal.

During recent years, following intractable crop protection problems arising from the misuse of insecticides, crop protection has been dominated by the concept of Integrated Pest Management (IPM). Essential to IPM is the Economic Threshold, defined as 'the density at which control measures should be applied to prevent an increasing pest population from reaching the Economic Injury Level' (Stein, Smith, Bosch & Hagen, 1959). IPM is thus concerned with managing, or regulating pest populations at least during the time when the crop is liable to be damaged.

It must be recognized, however, that although some insect pests are resident in the crop and the economic injury level is attained by multiplication within the crop, most crop pests are highly mobile. Southwood (1962) assembled massive data which showed convincingly that the level of migratory movement, which, by definition, was away from the occupied habitat, was a function of the degree of impermanence of the habitat. He pointed out (Southwood, 1971) that 'most arable crop plants are derived from ruderals, plants that are colonizers of bare ground, and occur early in the botanical succession. These are temporary habitats, and thus many pests of arable crops have, from their evolutionary history, high inherent levels of migratory activity'.

The economic injury level is achieved when the pest population density in a crop grows to a critical level, and population increase occurs when natality exceeds mortality plus emigration. However, natality is a function of numbers and reproductive capacity and both are related to the migratory activity of the species. Thus economic thresholds, which to be of value must be capable of making short-term predictions of population changes, are meaningless unless they take into account the effect on population change of immigration and emigration. Moreover, some pests colonize crops in numbers which themselves exceed the economic injury level.

Integrated Pest Management therefore demands that pest control is based on an adequate understanding of the contribution of migration to population change. In this sense most crop pests, though not migratory as defined by Kennedy (1961), must be studied for the purpose of control as if they were, inasmuch as they may breed on plants or in fields other than those where they themselves are born.

Migration and crop protection

Insect movements, as first expounded by Southwood and Kennedy, are generally accepted as being of two types, namely, trivial and migratory, the former being concerned with such behavioural compulsions as feeding, mating and shelter, whilst during the latter, insects fly persistently and are not distracted by these behavioural responses. A nice example of this is provided by Schaefer's (1976) radar observations in Sudan Gezira (Fig. 1) on the flight activity of the grasshopper *Aiolopus simulatrix*. During periods when winds were from the south the numbers of airborne grasshoppers observed by radar followed the same pattern as those caught at light traps operated by the late N. Russell–Smith (unpublished). When the winds changed to northerly, the grasshoppers were little distracted by light but took off and flew persistently with the wind. It is difficult to avoid the conclusion that this is adaptive behaviour which permits the species to reach its hibernation and aestivation areas located in the deeply cracking clay soils of the higher rainfall, tall-grass areas lying to the south (Joyce, 1952).

Students of migratory movements, which, by definition, take animals away from a habitat, have been usually concerned with exodus. On the contrary, from the crop protection viewpoint, immigration is usually more important than emigration. We are here concerned with the pattern of arrival of immigrants and with their reproductive capacity.

There appears to be a strong relationship between migratory activity and reproductive potential. Thus Dingle (1974) describes three species of the cotton stainer bug *Dysdercus* in East Africa (Fig. 2). *D. fasciatus* feeds on the fallen fruits of the baobab or tebeldi tree (*Adansonia digitata*) which are of seasonal occurrence, constituting an unstable habitat, unreliable in space and duration. *D. fasciatus* exploits this by having a high population growth rate, r, of 0.0939 per

Fig. 1. Day-to-day variation in airborne density of Sudan grasshoppers during the take-off period. Hollow circles, radar; filled circles, trap; crosses, temperature. Radma Sudan Gezira, 1971. (From Schaefer, 1976.)

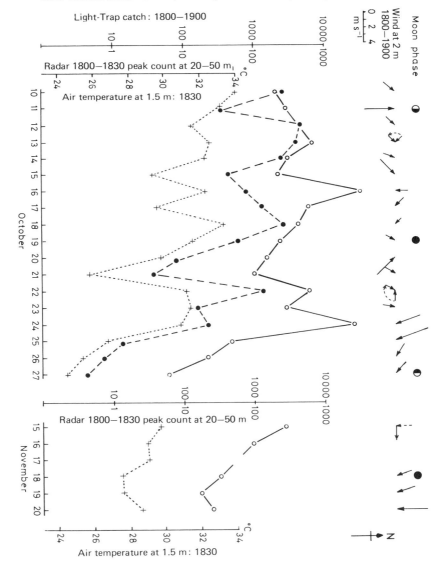

individual per day. On the other hand, *D. superstitiosus*, feeding on a wide variety of annuals and perennials, has a more stable habitat and can maintain its numbers with *r* at 0.0616. *D. nigrofasciatus* is intermediate, being confined for feeding to a limited number of plant species, all Malvales. Since damage to the cotton crop results from multiplication within it of initial immigrants, *D. fasciatus* is the most serious pest where several *Dysdercus* species are involved. Thus in Zambia the predominant invader is *D. superstitiosus* but *D. fasciatus* is more numerous towards the end of the season when bolls are damaged (Pearson & Maxwell–Darling, 1958).

The reproductive rate, *r*, is of course, not solely determined by the migratory activity of the pest and may vary markedly within a species. Thus Joyce (1961) found that the reproductive rates on cotton of the jassid *Empoasca lybica* de Berg and the whitefly *Bemisia tabaci* Gen. varied from season to season in association with the nitrogen content of the young leaves which they colonized, nitrogen content being predictable from the rains which fell six weeks before planting. In the case of *E. lybica* these differences in reproductive rates persisted even after the initial immigrants had been killed by DDT (*p,p'*-dichlorodiphenyl trichloroethane) spray.

In many cases economic threshold levels are irrelevant, whether or not the

Fig. 2. Reproductive values for three species of cotton stainer bugs *Dysdercus* in East Africa. The reproductive value indicates the expected contribution of an individual of specified age to future population growth, expressed as a multiple of its value at birth. Continuous line, *D. nigrofasciatus*; dashed line, *D. fasciatus*; dash-dot line, *D. superstitiosus*. First arrow, eclosion; second arrow, flight. (From Dingle, 1974.)

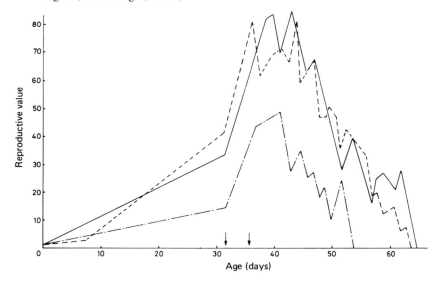

insects are considered migrant, because they invade the crop at densities which exceed the economic injury level. Locusts and grasshoppers are of economic significance clearly on this account, a typical swarm of desert locusts containing 50×10^6 individuals per km^2, weighing over 100 tonnes and eating their own weight of food per day.

Similar numbers of grasshoppers, such as *A. simulatrix* and *Oedaleus senegalensis* invade sorghum crops in the rainlands of Africa north of the equator, having bred on uncultivated land which might have been hundreds of kilometres from the crops they destroy. In fact Gunn & Rainey (1979) confined their Royal Society Symposium on Migrant Pests to those species 'that require control before they enter the area at risk'.

From the point of view of crop protection, this restriction on the definition of migrant pests is inadequate, because economic injury can result from the invasion of or accumulation within a crop of large numbers of the pest species from distant sources where it would be impractical or uneconomic to conduct control. Thus *E. lybica* and *B. tabaci* have been recorded on at least three occasions during the past two decades, colonizing cotton in the Sudan Gezira at damaging levels within ten days of its germination. The low density of host plants of *E. lybica*, on the one hand, and the vast number of those of *B. tabaci* on the other, on which these invaders had bred, preclude the possibility of destroying the pests before they enter the area at risk. Yet both species must be considered, for the purpose of control, as migrants which invade the crop in numbers exceeding the economic threshold.

Similarly, the noctuid *Heliothis armigera,* the damaging population of which in the Sudan Gezira breeds on host plants which include crops such as groundnuts and sorghum, invades the cotton in numbers in excess of the economic injury level repeatedly during September and October when the crop is most vulnerable. Destruction of these insects in their source areas is likely to be more expensive and more ecologically harmful than their destruction in the crop.

Migration considered as moving away from a habitat (and consequently to another habitat, which may be a crop), is not the only aspect of insect flight which is of importance in the development of a crop protection strategy. We are also concerned with redistribution which may be achieved by trivial movements. *H. armigera* typically lays single eggs, oviposition frequently alternating with flight. Thus a single female laying up to 100 eggs per night can travel several kilometres and provide infestations in several fields.

If IPM involves regulating populations it is necessary to have a clear idea of what constitutes the population to which control must be applied. Taylor & Taylor's concept (1977) is of special importance in this respect. They state 'all populations are spatially fluid in some measure because movement is a fundamental biological response to adversity. Nor is population density distributed uniformly in space. A valid concept of population must encompass settled species, like the red grouse, to which current models appear to be most suited, and the highly volatile species, like aphids, for which local stability in time seems almost irrelevant because of the persistence of change of place'.

A nice example of the constant redistribution of individuals composing a population is shown in Figs. 3 and 4, for which I am indebted to H. H. Coutts. They show the distribution of adult whitefly in a 4-ha field of cotton in the Sudan Gezira divided into 24 plots to which have been applied six insecticide treatments, each replicated four times. Isopleths have been drawn to show the distribution of whitefly density. It will be seen that neither the pre- nor post-spray distributions of densities are related either to each other or to the treatments applied.

Taylor & Taylor consider it necessary to treat the anatomy of a real population as being in three dimensions, latitude × longitude through time but having an internal reticulate structure analogous to that of a fern stele (Fig. 5).

It is in this context that the radar observations of Schaefer (1976, and unpublished) assume great importance in pest management. Population redistribution by flight has long been recognized, but it is usually considered that, by and large, emigration and immigration balance so that their effects on population change can be neglected. This is incorrect, not only because flight habits, in terms of time of flight, rate of climb, navigation procedures and the characteristics of the

Fig. 3. The distribution of whitefly in a 4-ha field of cotton in the Sudan Gezira. The field is divided into 24 plots. The figure in each plot shows the number of whitefly on a sample of 100 leaves taken from each plot. The 'contour lines' (isopleths) show the overall distribution of densities within the field. (From H. H. Coutts, unpublished.)

Pre-spray 10 January 1973

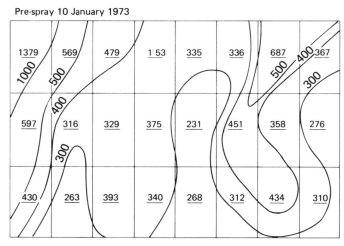

air in which species or individuals of a species choose to fly, are infinitely variable, but also because the air which they employ is structured. In consequence, some individuals of a population or some species of an ecosystem become grouped together in concentrations and some are dispersed. Some are deposited or alight in high densities, and some are transported away from the area at risk. Those individuals engaged in trivial flight may be transported and congregated in a different way from those engaged in migratory flight, but certain wind systems may affect both groups equally. Thus, over the Sudan Gezira, the Intertropical Front, the concentrating effect of which is weak (for example, doubling the area-density in 2–3 h, Rainey, 1976), can be expected to change the density distribution of only those insects whose flight is prolonged over several hours. On the contrary, the cold outflows from convectional storms which typically sweep over the Gezira just after sunset (when the numbers of insects engaged in both trivial and migratory flights are maximal), have a capability of doubling the area-density of airborne insects in about 2–3 min (Schaefer, 1976).

Such redistribution and concentration of pest populations have been studied in the Sudan Gezira by M. J. Haggis (unpublished), who has demonstrated that following the passage of convectional storms in the evening, a higher incidence of oviposition by *H. armigera* was found the following morning around the perimeter of the storm than along the track the storm had traversed. The late N. Russell–Smith found that such increases in the incidence of eggs of *H. armigera* were accompanied by increases in the incidence of *B. tabaci,* which occur in

Fig. 4. Whitefly densities in the same cotton field as in Fig. 3, after each of the 24 plots had been sprayed with one of six different insecticides. (From H. H. Coutts, unpublished.)

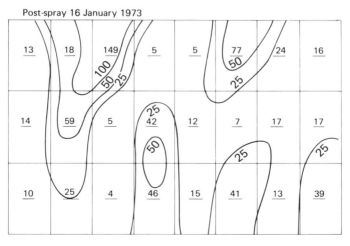

Post-spray 16 January 1973

high densities in the lower 50 m of airspace at this time (Rainey, 1976, and unpublished data).

Crops liable to damage by migrant pests in the sense used by Rainey cannot be protected by standard crop protection methods which, as far as chemical control is concerned, are based on the strategy of maintaining the crop as an environment lethal to the pest species. On the contrary, protecting crops against migrant pests involves identifying from amongst the total potentially damaging population that fraction which is actually the cause of crop loss, and developing methods to destroy or otherwise regulate those numbers. Control of migrant pests must be rooted in ecology and not agronomy.

Consideration of the dynamic nature of insect populations and the influence of both migratory and trivial flight on population change, leads inescapably to the conclusion expressed forcibly by Rabb (1970) that 'pest management deals primarily with populations, communities and eco-systems. Thus the basic biologi-

Fig. 5. Conceptual model for population anatomy based on Rostowzew's drawing of the stelar structure of the adder's tongue fern. (From Taylor & Taylor, 1977.)

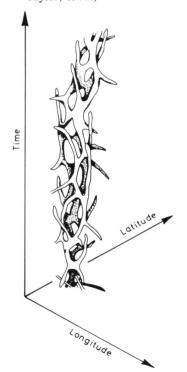

cal discipline involved is unequivocally ecology, and there must be no fuzzy thinking on this fact'.

I have described elsewhere (Joyce, 1973) the concept of Synoptic Survey, designed to determine the spatial and temporal dimensions of that fraction of the total potentially damaging population which occurs at unacceptable levels, and the need for such survey to be followed by the quasi-synchronized application of a regulator by a target-specific method, which ensures that the bulk of the regulator released into the environment is collected by a strictly defined target surface and the minimum by non-target surfaces. These concepts of Synoptic Survey and Synchronized Control, first developed for indisputably migrant pests such as the desert locust, have been found to have more general application to the management of insect pest species whose mobility has been invariably underestimated or overlooked.

Synoptic survey

Desert locusts in eastern Africa

The aim of synoptic, or more accurately, quasi-synoptic, survey is to obtain a picture of the distribution of a potentially damaging pest population at one time, and compare it with that at other times, so as to determine the location of the pest in actually damaging numbers or to forecast the arrival in crops to be protected of pests in numbers or in densities which can cause damage. The most intensive use of synoptic survey has been in locust and grasshopper control where damaging numbers arise outside the area at risk. Desert locust control has been made possible by the help and coordination provided by the Food and Agricultural Organization of the United Nations, which permits surveys to be conducted quasi-synoptically over the 30 million km² in sixty different countries where desert locusts may occur. Biogeographic analysis of the survey reports, pioneered by Rainey and his colleague, Z. Waloff, provides forecasts of the probability of invasion of a country as a result of breeding in other countries, as well as giving some indication of the likely scale of the invasion.

The objective of synoptic survey, as of science in general, is prediction but prediction in terms of likelihood of occurrence. Synoptic survey in locust control is described elsewhere (particularly in various Anti-Locust Research Centre Memoirs) but one mention is made here. Fig. 6 is an example of an aerial sampling scheme for bands of desert locust hoppers in the Somali Peninsula, where the strongly negative-binomial distribution of populations could be regarded as log-normal. The scheme was designed to provide a 98% chance of discovery of all lots (10-min squares measuring approximately 400 km²) which were infested to an intolerable level (selected as one band per square kilometre), and a 5% chance that a lot rejected for control as uninfested was in fact infested beyond the tolerable level. Fig. 7 shows the probability of swarm formation when control measures had been applied to each lot designated infested. In practice, an aerial traverse had to be made to search 10% of the area and the discovery of a single band in this sample provided the security designed into the sampling scheme.

Table 1 summarizes some of the results of this sampling scheme.

Cotton bollworm in the Sudan Gezira

Some 250 000 ha of cotton are grown in about 1 000 000 ha of crop area irrigated by the Blue Nile in the Sudan Gezira under a single management, the Sudan Gezira Board. Larvae of the noctuid *H. armigera,* which derive from moths bred on groundnuts, vegetables and irrigated sorghum as well as rain-grown sorghum inside and outside the Gezira, destroy developing fruit of the

Fig. 6. Characteristics of a scheme for the aerial sampling of desert locust hopper bands. The area is divided into 400-km² lots, and 10% of each lot is inspected. The sampling scheme is based on the contiguous nature of the distribution of hopper bands, such that the discovery of a single hopper band in the sample provides (i) a 98% chance that the lot is in fact infested beyond a tolerable level (over 3% of the area infested by hoppers), and (ii) a 5% chance that the lot accepted as uninfested was in fact infested beyond the tolerable level. Notice that the great majority of the heavily infested lots are discovered by the sampling scheme, and also (Fig. 7) that the majority of escapes come from the relatively few hopper bands in the lots which are lightly infested. The scheme was tested with apparently satisfactory results in the Somali Republic in December 1960, in an area considered uninfested by ground teams.

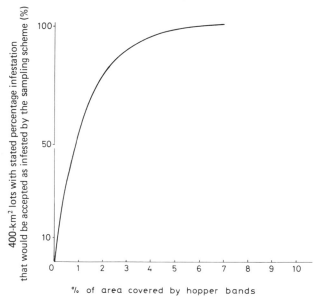

cotton plant between 30 and 80 days after the cotton is planted, as a result of repeated invasions of the cotton fields. The economic injury level is reached when there are about 10 000 larvae amongst a plant population of about 75 000 per ha. This density of larvae could be generated in 2–3 days after an incursion of 50–100 ovipositing moths per ha. Within 10–30 min of eclosion the larvae begin to damage the plant's growing points or the primordial buds. If damage is to be minimized, it is necessary to prevent the moths from invading the cotton, or destroy the eggs or the larvae immediately following eclosion. In any case, it is necessary to know daily the distribution of moths over the entire area at risk, so as to locate the areas where they exist at intolerable densities. The time dimension of quasi-synoptic survey in this case is one day, in contrast with up to 30 days for desert locust hoppers, and less than one day for locust swarms. A knowledge of the scale of movements, in particular the flight activity, of a pest is essential for establishing the time dimension of quasi-synoptic survey.

Laboratory studies (Hackett & Gatehouse, 1978) confirmed Schaefer's (1976) radar and Topper's (unpublished) uninstrumented field observations that the majority of adult *H. armigera* (over 85% in the laboratory) were airborne within one hour after sunset and engaged in flights (in the laboratory of 5–10 min),

Fig. 7. Aerial sampling of desert locust hopper bands: escapes from the sampling scheme described in the caption to Fig. 6.

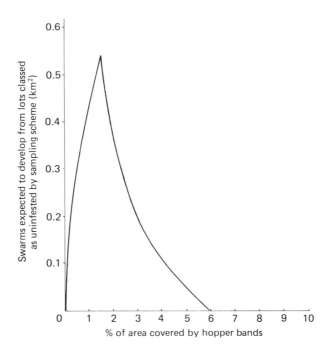

which would transport them at least one kilometre (that is, across several cotton fields). The remaining 15% of the moths flew in the laboratory for longer periods (>30 min). Flight was most sustained amongst sexually immature and unfed adults 2–4 days old, and decreased after mating, which was followed the next day by oviposition. The adults live 8–13 days (Balla, 1970). Thus, in a population of adult *H. armigera,* not more than 4% are likely to engage in flight which would transport them much more than one kilometre. Schaefer (1976) with his radar observations confirms this, only a small fraction of the echoes modulated at the wing beat frequency of *armigera* being recorded above Taylor's (1958) 'boundary layer', though these persisted for more than six hours and would be likely to have carried the moths outside the Gezira. The flight of most adults is within the lower 5–10 m and most of it trivial.

In practice it was found that the most cost-effective method of survey was to use the occurrence of eggs on cotton as an indication of the recent presence of ovipositing adults, so that a random sample had to be taken daily from the whole area at risk, and be repeated each third to fourth day (the incubation of eggs occupying 2–3 days at this time of the year). The problem of daily survey of hundreds of thousands of hectares of cotton was studied by Joyce and Russell‑Smith. It was found that the economic injury level occurred when only 15% of

Table 1. *Production of swarms of desert locusts in the Somali Republic as predicted by aerial survey of hopper bands (Joyce, 1963)[a]*

Breeding season	Expected	Actual
Southern region		
Oct–Dec 1960	650	520[b]
Northern region		
May–June 1961		
Borama	13	25
Duruksi	65	Consistent[c]
Las Anod	2	0
Southern region		
Oct–Dec 1961	250	220[b]
Northern region		
May–June 1962	25	13

[a] Values are in km^2.
[b] Estimated from aerial reconnaissance.
[c] No detailed aerial reconnaissance possible but subsequent information was consistent.

the plants were infested. In order to obtain an estimate of the mean egg infestation with a 20% S.E. in an administrative block, occupying some 200 km² and containing 2000–3000 ha of cotton, one sample of 60–100 plants had to be taken from 5–10 standard, 40-ha fields, that is, from 2–5% of the fields, or less than one plant per million.

Typical results of these surveys in southern Gezira are given in Figs. 8 and 9, in which it will be seen that, on occasions, over 2500 km² of crop area, containing over 50 000 ha of cotton could be infested in any 2–3-day period in excess of the economic injury level. Not only was it necessary to destroy the larvae before they caused damage, but also it was desirable to destroy the moths responsible for the eggs before they moved to cotton in which the infestation was still at an acceptable level.

Thus synoptic survey defines the scale upon which action must be taken to minimize crop loss. The scale of this survey in respect of both space and time is a function of the pest's distribution and redistribution, which are determined by its flight behaviour.

Methods of synoptic survey

Radar, an invaluable tool for studying insect flight, suffers from its inability to provide specific determination of the echoes. When only one species dominates the airborne fauna, as with the spruce budworm *Choristoneura fumiferana* Clem., this limitation is unimportant. In many cases, however, specific identification can be assured only by examination of captured specimens. Light traps are often useful, and their employment in survey for the rice stemborer *Tryporyza* spp. has been described elsewhere (Joyce, 1976). Perhaps the most extensive grid of light traps is that established by the late E. S. Brown for monitoring and forecasting the probable movement of the African armyworm *Spodoptera exempta* Wlk in eastern Africa. The interpretation of the light-trap data and their value in forecasting was reviewed by Betts (1976). Haggis (1971) using light-trap data, showed the effect of the wind shifts associated with the African Rift Convergence Zone in concentrating airborne populations prior to outbreaks. Rainey (1973, 1976) described several meteorological systems in eastern Africa which could explain the variability of light-trap catches recorded in the grid and subsequent outbreaks of larvae.

Suction traps, pioneered by Johnson, are used extensively in the United Kingdom, and their value in synoptic survey and forecasting have been described, for example by Taylor (1973).

Like light traps, suction traps suffer from the difficulty in interpreting their catches quantitatively, since these represent a mean over a period of time, rarely less than one hour, and provide information only on densities in the lower levels of the atmosphere, rarely more than tens of metres above ground level. This

Fig. 8. Distribution of eggs per 100 plants of the cotton bollworm *Heliothis armigera* on cotton in the Sudan Gezira as determined by sampling during a particular 3-day period, 22–24 September 1975. The area is divided into administrative blocks, designated by the integer figures (the southernmost block is number 95, for example). The decimal figures in each block show the numbers of eggs per 100 plants. The letter S in a block indicates that that block was sprayed with insecticide during the period of the survey, in response to counts exceeding the economic injury level in the previous 3-day period.

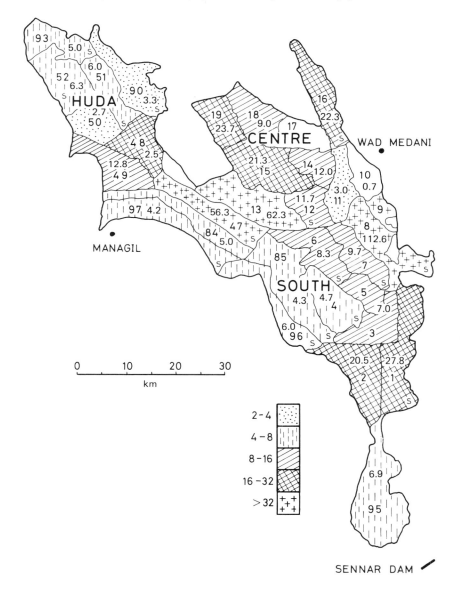

limitation need not, however, detract from their value in synoptic survey where we are more concerned with the damaging or potentially damaging fraction of the population than with migration *per se*.

A promising tool for synoptic survey is the use of pheromones. Lewis & Stur-

Fig. 9. Distribution of *Heliothis armigera* eggs in the Sudan Gezira in the 3-day period following that shown in Fig. 8, i.e. 25–27 September 1975. NA, not applicable.

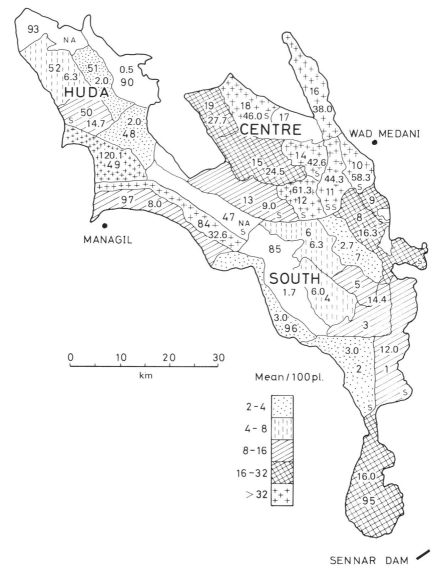

geon (1978) found that when males of the pea moth *Cydia nigricana* were first found in traps baited with a sex-attractant, females could be assumed to arrive in 0–3 days, so that eggs were likely to start hatching in 10–13 days under conditions near Rothamsted. Campion, Bettany, McGinnigle & Taylor (1977), on the other hand, were unable to establish any correlation between the catches of *Spodoptera littoralis* in Crete in traps baited with a sex pheromone and subsequent oviposition on potatoes. Nevertheless, the method provided information on the incidence of moths in potentially damaging numbers.

All these methods are tools by which information can be obtained on the incidence of a pest population in damaging or potentially damaging numbers at any one time throughout the current range of its distribution.

Synchronized control

Synchronized control, more accurately quasi-synchronized, describes the application of a regulator as speedily as possible to the entire pest population, where present in numbers which place at risk the crop to be protected. Since the locations occupied by the pest are, in the case of mobile or migrant pests, usually large and not defined by field or farm boundaries, synchronized control demands not only the use of target-specific methods of application if it is to be economic and speedy, but also the capability of operating on a scale dictated by the pest's ecology and behaviour, rather than by artificial constraints imposed by agronomic practice.

This principle is nicely illustrated by the aerial spraying of the insecticide, phosphamidon, in Indonesia against the rice stemborers, chiefly *T. incertulas,* at ultra-low-volume (ULV) rates which permitted a rate of work of up to 1000 ha h^{-1} (Table 2). The larger the area the more effective was the crop protection achieved, presumably because large areas were reinfested more slowly than small ones.

Again, in the Sudan Gezira, the quasi-synchronized application of the insec-

Table 2. *Control of rice stemborers in Java (from Singh, unpublished – see Joyce, 1976)*

Method of treatment	Size of area sprayed synchronously (ha)	Moth catch at light trap (%)	Paddy yield (%)
Aerial ULV[a]	5000	29	195
Aerial ULV	1000	63	140
Aerial ULV	500	60	141
By individual farmers	(0.2–2.0)	100	100

[a]ULV, ultra-low-volume rates.

ticide, Nuvacron, against *H. armigera* on a scale determined by synoptic surveys, again at ULV rates which permitted very high rates of work, have resulted
in impressive and consistently improved yields compared with those obtained
when cotton is treated, usually with more insecticide, by conventional aerial
spraying methods, timed in accordance with surveys for economic thresholds of
larval infestation not organized on a synoptic basis. Table 3 illustrates these
differences and Fig. 10 the pattern of cotton picking. Not only were the total
yields bigger, but also the proportion of the yield from the earliest picks, which
provide the best grades of cotton, were proportionately and absolutely greater.

Target-specific methods of pesticide application
    This is not the place to describe in any depth target-specific methods of
pesticide application. The subject is introduced to emphasize that treatment of
the control of pests as an ecological, rather than an agronomic problem, concerned with the management of a dynamic pest population, directs attention
towards the insect rather than the crop. Conventional use of insecticides is usually concerned with crop spraying, that is, to treat the crop so that it becomes an
environment toxic to the pest species, the 'toxic carpet' of Morton (1979). This
is only one route to insect pest management, and, frequently, a far better route
is to direct the regulator, which today is almost invariably an insecticide, at the
pest species, where it exists in intolerable numbers, by the most direct route and
under conditions where contamination of non-target surfaces may be minimized.
Insecticides have been most efficiently employed against flying locust swarms,
when there has been evidence that as much as 50% of the volume applied has
been collected by the target insect (Rainey, 1976) and more than 5% of the toxic
doses applied accounted for by locust corpses. This contrasts with the use of

Table 3. *Yields of seed cotton in the Sudan Gezira from various administrative
groups as a percentage of the yield of the whole Gezira*

Season	Groups subjected to synoptic survey and synchronized control			Groups treated by conventional methods	
	South	Centre	Managil	Old Gezira	Managil
1973/74	135	–	–	97	–
1974/75	134	121	–	87	–
1975/76	111	93	73	54	46
1976/77	115	121	92	86	66
Long-term mean 1961/2–1970/1	106	105	99	99	97

insecticides in conventional crop spraying quoted by Graham-Bryce (1977), where efficiency in these terms was of the order of 0.02%.

Schaefer (1976, and in press) in his radar observations on the spruce budworm moth found that night after night a high proportion of the total moth population occurred at densities of about 3 per 1000 m³ up to a height of 200 m. These are sprayable targets. Lawson (unpublished) studied the dispersal of a cloud of droplets in the air in which the moth flew. The spray was released at 6 l km⁻¹ about 250 m above ground level, and was composed mainly of droplets 30–60 μm diameter. These droplets remained airborne for up to one hour after release at sufficiently high density to permit a moth flying through the cloud to collect several LD_{95} doses. This expectation was confirmed by Solang Uk (unpub-

Fig. 10. Effect of synchronized control on cotton yields in the Sudan Gezira. The graphs show accumulated monthly yields of seed cotton (*Gossypium barbadense*) during the 1975/76 season from areas in which pest control was based on synoptic survey and synchronized control (continuous lines) compared with conventional practice (dashed lines).

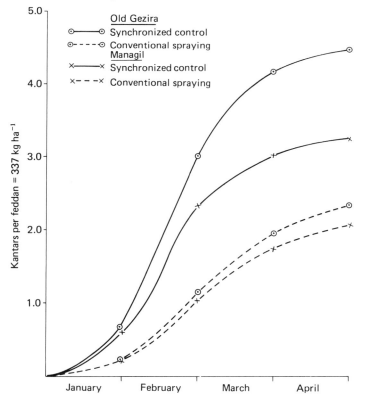

lished), who measured the amount of chemical collected by moths which had traversed such a spray curtain. At a wind speed of 7 m s^{-1} this meant that 2500 ha min^{-1} could be sprayed from a light aircraft and the pest population destroyed with about 2 g ha^{-1} active ingredient. This compares with the annual spraying of millions of hectares of forest in eastern Canada against larvae of the same species with about 100 times the amount of active ingredient and at a work rate 50 times slower.

In Sudan Gezira adults of *H. armigera,* which can generate larvae at economic injury levels at densities of about 50 moths ha^{-1}, are too few to present a suitable airborne target. Nevertheless, the tactic of applying an insecticide to all the cotton where eggs indicate the recent presence of adults at unacceptable levels, has been found to bring oviposition to an immediate halt by destruction of the moths responsible. This is achieved by at least 3 routes, namely, by direct contact, by movement of insecticide to the extra-floral nectaries where moths feed, and by residual contact action on contaminated foliage. None of these routes is effective for more than 36 h, so there is no lasting effect on other insects associated with the crop, many of them predators and parasites.

Conclusions

In crop protection we are concerned with

(1)   immigration into (and emigration from) the crop,
(2)   the reproductive capacity of immigrants, and
(3)   redistribution of individuals,

in order to prevent infestations occurring at economic injury levels. The scale of our pest surveys and control measures are determined by these three factors – and the factors themselves are a function of the migratory status of the pest.

It is now becoming apparent that target-specific methods of pesticide application, which are essential for the chemical control of migrant pests, are also essential for the control of pests not normally considered migrant, if the unwanted effects of the misuse of chemicals are to be avoided.

Only in the cases of those pests, the habits of which result in the completion of colonization of a crop before the emergence of damaging numbers, can field boundaries determine the scale on which control must be applied. It is only when control can be applied on a scale determined by field boundaries that standard cost–benefit analysis of the operation can be employed. Such pests of course exist, but it is becoming all too clear that the migratory activity of most crop pests has been grossly underestimated, and that insufficient attention has been directed at the control problems which this activity generates.

Inasmuch as arable crops are essentially temporary and impermanent habitats, most pests associated with them show a high degree of migratory activity. The principles of synoptic survey and synchronized control, developed for migrant pests, thus have a wide application to pests which are mobile.

One reason for this failure to take into account a fundamental activity of all living organisms is not just the difficulties involved in studying population dynamics, taking into account immigration, emigration and redistribution, but that most research on the control of crop pests is conducted on small plots, laid out to satisfy the statistical requirements of agronomic techniques. Under these conditions the problems of pest mobility are considered to be absent, or that they affect all plots equally. It is only when insect control has to be undertaken in large-scale commercial practice, that is, in the real world, that problems of pest mobility become unavoidable. It is significant that the new data on insect mobility and crop protection reported here from Indonesia, Sudan and Canada, have been acquired because they have been necessary for a commercial company, CIBA–GEIGY, to fulfil contractual obligations of providing an insect-control service. I am grateful to the Company for the opportunity it gave me to participate in and lead much of this research, which I believe has been of real benefit to science. I am also grateful to the Company for permission to quote from some of its unpublished reports.

References

Balla, A. N. (1970;. American bollworm in the Gezira and Managil. In *Cotton Growth in the Gezira Environment*, ed. M. A. Siddig & L. C. Hughes, pp. 281–92. Wad Medani: Agricultural Research Corporation of Sudan.

Betts, E. (1976). Forecasting infestations of tropical migrant pests: The desert locust and the African armyworm. In *Insect Flight*, ed. R. C. Rainey, pp. 113–34. Oxford: Blackwell Scientific Publications.

Campion, D. G., Bettany, B. W., McGinnigle, J. B. & Taylor, L. R. (1977). The distribution and migration of *Spodoptera littoralis* (Boisduval) (Lepidoptera: Noctuidae), in relation to meteorology on Cyprus, interpreted from maps of pheromone trap samples. *Bulletin of Entomological Research,* **67,** 501–22.

Dingle, H. (1974). The experimental analysis of migration and life-history strategies in insects. In *Experimental Analysis of Insect Behaviour*, ed. L. Barton–Browne, pp. 329–42. New York: Springer-Verlag.

Graham-Bryce, I. J. (1977). Crop protection: a consideration of the effectiveness and disadvantages of current methods and of the scope for improvement. *Philosophical Transactions of the Royal Society of London* B, **281,** 163–79.

Gunn, D. L. & Rainey, R. C. (1979). Strategy and tactics of control of migrant pests. *Philosophical Transactions of the Royal Society of London,* **287,** 245–488.

Hackett, D. & Gatehouse, A. G. (1977). Larval development of *Heliothis armigera* Hb. on groundnuts, sorghum and cotton in the Sudan Gezira. In the *Report of Third Seminar on the Strategy for Cotton Pest Control in the Sudan*. Basle: CIBA–GEIGY.

Haggis, M. J. (1971). Light trap catches of *Spodoptera exempta* (Walk.) in relation to wind direction. *East African Agricultural and Forestry Journal*, **37**, 100–8.

Joyce, R. J. V. (1952). Ecology of grasshoppers in East Central Sudan. *Anti-Locust Bulletin*, **11**, 103 pp.

Joyce, R. J. V. (1961). Some factors affecting numbers of *Empoasca lybica* (de Berg) infecting cotton in the Sudan Gezira. *Bulletin of Entomological Research*, **52** (1), 191–232.

Joyce, R. J. V. (1963). *Some Assessment Methods. United Nations Development Programme Desert Locust Project.* Symposium No. 3. (unpublished typescript).

Joyce, R. J. V. (1973). Insect mobility and the philosophy of crop protection with reference to the Sudan Gezira. *Pesticide Articles and News Summaries*, **19**, 62–70.

Joyce, R. J. V. (1976). Insect flight in relation to pest control. In *Insect Flight*, ed. R. C. Rainey, pp. 135–155. Oxford: Blackwell Scientific Publications.

Kennedy, J. S. (1961). A turning point in the study of insect migration. *Nature*, **189**, 785–91.

Lewis, T. & Sturgeon, D. M. (1978). Early warning of egg hatching in pea moth (*Cydia nigricana*). *Annals of Applied Biology*, **88** (2), 199–210.

Morton, N. (1979). Synthetic pyrethroids on cotton: a spray application strategy. *Outlook on Agriculture*, **10** (2), 71–9.

Pearson, E. O. & Maxwell-Darling, R. C. (1958) *The Insect Pests of Cotton in Tropical Africa.* London: The Commonwealth Institute of Entomology.

Rabb, R. L. (1970). In *Concepts of Pest Management*, ed. R. L. Rabb & F. E. Guthrie. North Carolina State University.

Rainey, R. C. (1973). Airborne pests and the atmospheric environment. *Weather*, **28** (6), 224–39.

Rainey, R. C. (1976). Flight behaviour and features of the atmospheric environment. In *Insect Flight*, ed. R. C. Rainey, pp. 75–112. Oxford: Blackwell Scientific Publications.

Schaefer, G. W. (1976). Radar observations on insect flight. In *Insect Flight*, ed. R. C. Rainey, pp. 157–197. Oxford: Blackwell Scientific Publications.

Southwood, T. R. E. (1962). Migration of terrestrial arthropods in relation to habitat. *Biological Reviews*, **37**, 127–214.

Southwood, T. R. E. (1971). The role and measurement of migration in the population system of an insect pest. *Tropical Science*, **13** (4), 275–8.

Stein, V. M., Smith, R. F., Bosch, R. & Hagen, K. S. (1959). The integrated control concept. *Helgardia*, **29** (2), p. 81.

Taylor, L. R. (1958). Aphid dispersal and diurnal periodicity. *Proceedings of the Linnean Society of London*, **169**, 67–73.

Taylor, L. R. & Taylor, R. A. J. (1977). Aggregation, migration and population mechanics. *Nature*, **265**, 415–21.

Taylor, L. R. (1973). Monitor surveying for migrant insect pests. *Outlook on Agriculture*, **7** (3), 109–16.

G.V.T.MATTHEWS

The conservation of migratory birds

The need for the rational conservation of sedentary game was a relatively easy concept to grasp, and thus developed early in history. For if too many sedentary animals were killed, the diminution in the population rapidly became obvious. The balance could be redressed by killing fewer animals, protecting breeding stocks from predators and disturbance, supplementing the wild population with artificially reared animals and establishing sanctuary areas. The habitat of the animals could be improved by management and food sometimes supplied directly during critical periods. The results of these conservation measures, relatively easy to enforce in authoritarian societies, became readily apparent to the hunters, so encouraging their efforts.

Where migratory animals, especially birds, were concerned, the situation was much less easy for the exploiter to comprehend. Indeed such birds were as manna from heaven, to be taken ruthlessly when they appeared, without thought for how they would be replaced. Until the basic facts of migration had been understood, little could in any case be done to protect breeding stocks. In the fifteenth century it would no doubt have made sense to conserve barnacles in the hope of increasing the numbers of barnacle geese.

Another feature of migratory animals which makes it difficult for the individual hunter to grasp the need for conservation, is their tendency to concentrate in flocks in relatively few areas. The behaviour becomes exaggerated as more and more natural habitat is destroyed by man's activities. It is not easy to appreciate that a great mass of geese, filling the sky over a lake in southern Europe, is the product of tens of thousands of hectares in the far northern breeding grounds.

However, during the last century conservation measures with regard to migratory birds were initiated at state level in the United States. Thus in 1846 the very destructive spring-shooting of waterfowl was forbidden on Rhode Island. By 1880 there was some sort of legal protection of wildlife in all the American States and Territories. But commercial killing for plumes and meat was still getting out of hand and the migrating hordes of the passenger pigeon had been thrust towards extinction. Inter-state legislation was clearly needed and was obtainable through the federal structure of the United States. The Lacey Act was slipped through Congress in 1900 with remarkably little opposition. It made illegal the inter-state shipment of birds taken contrary to state law, and was the beginning of the end of market-hunting in North America. The Lacey Act also regulated the introduction of exotic species. This was necessitated by the deplorable tendency of homesick colonials to release birds from Europe in the new lands – often with disastrous results, as in New Zealand and Hawaii.

In Europe the conservation of migratory birds moved hesitantly. Inter-state legislation was particularly difficult because of the multiplicity of languages and of nationalistic barriers. The feudal system, with the landowners looking after their selfish interests, had provided an effective check on over-exploitation. But as the great estates began to break up, under social revolution and economic pressures, action at governmental level became more pressing. In Britain an Act for the Preservation of Sea-birds was passed in 1869, followed by an Act setting close seasons for all wild birds in 1880 (the centenary was celebrated by a special issue of stamps in 1980). Another ten Acts followed, until the Wild Ducks and Geese Protection Act of 1939. The legislators had tended to concentrate on their own favourite groups of birds and it was not until 1954 that the comprehensive Bird Protection Act (amended in 1967) was passed. This embodied the important concept of protecting all birds with the exception of named lists of huntable species (with close seasons) and of the so-called 'pest' species that impinge too markedly on man's interests, mainly in agriculture. Yet despite this close attention to the welfare of birds, Britain still lacks comprehensive wildlife protection legislation such as provided by the brilliantly drafted Irish Wildlife Act of 1976. The comprehensive-sounding Countryside and Wildlife Bill shortly to be placed before Parliament will not do much more than cobble together (and up-date) existing legislation.

Although an inter-state convention on bird protection in Europe was discussed in 1868 in Vienna, it was not until 1902 that 12 countries signed in Paris the International Convention for the Protection of Birds Useful to Agriculture. Such emphasis on 'good' birds may seem antiquated nowadays. However, with two wars intervening, it was not until 1950 that a more comprehensive Bird Protection Convention was signed by thirteen governments, also in Paris. This made the mistake of detailing protection measures in too much detail, with the result that several countries with otherwise good record in conservation, including Britain, have never ratified their signatures (i.e. accepted it into their national legislation). Although officially brought into action when sufficient countries had so ratified, the Paris Convention is generally regarded as something of a 'dead duck'.

Much more effective is the Directive on Bird Conservation accepted by the Ministers of the nine countries of the European Economic Community in April

1979. The Treaty of Rome requires that any Directive be incorporated in national legislations within two years, and any infringements thereafter will be dealt with by the International Court at the Hague. This legal backing makes the Directive far stronger than a Convention. Although it covers only a limited geographical area, Western Europe, this provides the wintering grounds of many species of birds breeding to the north and east. When the EEC is enlarged to incorporate Greece, Spain and Portugal, these three countries, at present conservationally backward, will immediately have to bring their bird protection laws into line with the Directive.

The EEC Directive incorporates the 'all-birds-protected-except' principle, but it goes further than the British in eschewing an automatic 'pest' list. Instead, derogations will have to be justified in the Commission of the EEC in respect of each species it is desired to control. The recognition that birds (and other animals) only become pests where man has altered the environment, for instance by establishing monocultures, is biologically pleasing.

The list of huntable species has been drastically pruned from 117 to 72, only 24 of which may be hunted throughout the entire Community. Furthermore only seven huntable species (those frequently propagated artificially, such as pheasants) may be sold throughout the community. The populations of another 19 species and the likely effect of marketing them are being urgently reviewed. It is probable that many of these will be excluded from the market-place, thereby following the lead established in North America more than sixty years ago. It certainly is unacceptable that private profit should be made by killing animals reared in countries thousands of kilometres away. Monetary gain is one of the major stimuli for the hunter to kill more birds than can be eaten by his family. Argument about protein requirements can really cut little ice, at least in Europe.

The Directive further seeks to prevent excessive kills by banning many methods of capture and kill that are indiscriminate and can lead to mass slaughter (nets, traps, poisons, automatic weapons and the like). These provisions should be especially relevant in the Mediterranean countries which have hitherto made substantial inroads into the migratory song-birds reaching that area from countries farther north.

Important as is the control of kill, the Directive's most far-sighted provisions are those requiring the setting aside of areas of habitat vital to vulnerable and threatened species, habitat which is used by migratory species for breeding, moulting and wintering, and as staging posts along their migration routes. Member states are required to pay especial attention to wetlands, the most easily threatened group of habitats, especially those of international importance. Appropriate steps must also be taken to avoid pollution or deterioration of the protected habitats. An important practical point is that the Directive provides for a Committee of Adaptation which will modify the provisions in the light of changed circumstances – and makes available the necessary funds for research and survey.

Such whole-hearted support for basic ecological principles is most heartening. Shakespeare, of course, said it all before – 'You take my life, when you do take the means whereby I live' (Merchant of Venice).

A certain amount of duplication with the EEC Directive has been brought about by the acceptance in Berne, in September 1979, of the Council of Europe's Convention on the 'Conservation of European Wildlife and Natural Habitats'. This has been signed by nineteen countries and thus has a potentially wider geographical coverage. But it has yet to be ratified by any state (five are required to 'trigger' it into action) and even then will not have the strong legal backing afforded to the Directive. The Convention seeks to cover all types of wildlife, not just birds, and to protect plants as well as animals. By that very token it cannot but be less precise in its injunctions. Unfortunately the 'all-species-protected-except' policy has been abandoned in favour of protected species lists, where omissions and mistakes will inevitably give rise to controversy. Special provisions are made for migratory species, to coordinate national efforts in conservation of habitats and in regulation of hunting.

Although the Council of Europe's Convention has, rightly, been called 'weak', it will at the very least focus attention on nature conservation in those countries that had hardly considered it. And even the more advanced countries need reminding that there is more to wildlife than just birds.

Countries outside the Council of Europe could, in theory, adhere to its Convention. In practice it is unlikely that the countries of Eastern Europe would be interested in doing so. However, another basically economic organization, one of the United Nations' offspring, the Economic Commission for Europe (ECE rather than EEC), convened a high-level meeting on the protection of the environment in September 1979. The ECE, since it covers all European countries without respect to politics, might perhaps provide a means whereby a formula for Europe-wide conservation could be achieved. It may be noticed in passing that the UN has set up Economic Commissions for other regions – Africa, South America, etc.

While Europe is only just emerging, we hope, into an era of continental conservation, North America has had long experience of such broad-based activity. In 1916 the United States and Canada (Britain acting on the latter's behalf) signed a Treaty on Migrating Birds. This was subsequently extended, in 1936, to Mexico. It firmly established the concept that migratory birds are not the property of any one state but are the shared responsibility of all the territories through which they pass. It also follows that they are a shared resource of these territories, and much subsequent activity has been concerned with ensuring an equitable distribution of any 'harvest'. This is a sportsman's harvest for, as we have seen, commercial exploitation. The list of birds harvested is limited and concerns mainly the ducks and geese. The shorebirds (waders) were long ago excluded from the hunting bag, as, indeed, they should be in Europe.

Since the 1940s, American waterfowl management has been based on the 'flyway' concept whereby the migratory populations are considered as moving southwards in four major streams, Atlantic, Mississippi, Central, and Pacific, deriving, in the main, from different parts of the northern breeding grounds. Breeding is seldom equally successful throughout the north, so the flyway system enables some fine-tuning to be attempted when deciding the harvest which can be taken in different places. The ultimate aim, management by breeding units, is being attempted in certain goose species.

Basically the waterfowl managers seek to establish the size of the breeding population in each major breeding ground, the state of the environment insofar as it affects breeding success (such as snow cover, groundwater) and, when possible, the production of young. These measurements involve regular and repeated sampling transects, usually from the air. From these data is estimated the size of the population that will come south in autumn. A further estimate is made of the harvestable surplus that could be taken without reducing the next year's breeding population below acceptable limits. This surplus is then allocated in proportion to the number of hunters and the time which the birds stay in each passage or wintering area. The share-out clearly involves a good deal of political manoevring between the various Provinces and States, only partially assisted by there being a common language. Translation of the allocation into dead ducks is achieved by an intricate, flexible system. The opening and closing of the shooting season can be varied, sometimes split. The bag limit – the number of birds which may be shot each day – can also be varied. There is, further, a points system designed to throw more attention on species with larger harvestable surpluses. They score low, whereas scarce species score high. Each hunter is allowed a total of so many points a day and is supposed to stop shooting when the limit is reached. In fact there are so many ways to 'fiddle' this points system, that it is of doubtful value. Nevertheless the flexible controls taken in concert have resulted in harvests of approximately the desired size.

For a long time it was assumed that, as hunting mortality was added to natural annual mortality, the hunting harvest thus directly affected subsequent size of a breeding population. Recent analyses here, however, lend strong support to the hypothesis that, below a certain threshold point, hunting mortality is compensated for by a lowering of natural mortality. Below that threshold the varying of hunting regulations will have little effect on subsequent population size; above the threshold, however, hunting could have a drastic effect.

It would now seem that for the opportunistic species of ducks, those reproducing in their first year, having large clutch sizes and short life spans (e.g. mallard, pintail, American wigeon), the threshold (40%) is well above the hunting mortality imposed in North America (but not necessarily in Europe). For less dynamic species with deferred maturity, small clutch size and long life spans, such as the diving ducks, the threshold (10%) is considerably below the hunting level imposed.

For the former the whole complex system, data collection and analysis, harvest allocation and regulation manipulation, would seem to have been a waste of time costing untold millions of dollars and man hours. For the latter varying regulations had to be abandoned as inadequate and seasons not opened at all (e.g. redhead and canvasback).

We can learn from others' mistakes and avoid the introduction of over-complicated management systems into Europe. Indeed such a system would be impractical. It is unlikely that a polyglot collection of states could agree on the sharing of a calculated 'harvest' in time for it to be harvested. It is unlikely that sufficient governmental money will be forthcoming to enable the detailed surveys and monitoring to be carried out. A good deal, however, can be achieved using volunteer observers, and the International Waterfowl Research Bureau based at Slimbridge does coordinate an annual midwinter wildfowl count, with reports from some 5 000 sites throughout Europe. It has also been able to gather some, much more dubious, data on the numbers of waterfowl killed. To give some idea of the comparative problems, North America has 100 million ducks and harvests 16.7 million of them; in Europe there are only 25 million ducks, yet 11 million (44%) are harvested.

A hitherto insurmountable block to data collecting in Europe has been the political inaccessibility of the main breeding grounds of 'our' waterfowl, the Soviet Union. However, the development of an Anglo-Soviet Environmental Programme, which is leading to an 'Anglo-Soviet Convention on Migratory Birds', may well open the curtains to some extent.

The Soviets have given the lead in intercontinental conservation measures by concluding Bird Migration Conventions with the USA, and also with Japan, covering the bird populations that migrate from East Siberia down the Pacific coasts. The USA also has a Migratory Birds Convention with Japan, which, in turn, has a similar treaty with Australia. The USSR/USA Convention, besides making the usual general undertakings to conserve shared stocks of migratory birds, also requires the listing and conservation of designated habitat areas.

It will be apparent – and it is certainly borne in upon the governmental bureaucrats who have to carry out the paper work and organize the necessary surveys and reports – that there is a plethora of migratory bird conventions, with some degree of overlapping. In an attempt to bring order into the system, work has been proceeding on an 'umbrella' Convention, following a Resolution of the

United Nations Conference on the Human Environment in Stockholm, 1972. In June 1979 the 'Convention on the Conservation of Migratory Species of Wild Animals' was concluded at Bonn, West Germany and signed by 44 countries (who will have to ratify these signatures in due course). Unfortunately several powerful countries with vast territories under their control such as Australia, Canada, the Soviet Union, and the USA refused to sign the Convention. This was mainly because the Convention, in seeking to be truly global in character, called for the inclusion of the Arctic, Antarctica and the Oceans within its scope. Unless these countries can be brought to modify their opposition, the Convention will be seriously weakened.

Be that as it may, Parties to the Convention are called upon to promote, cooperate in, or support research relating to migratory species and to endeavour to provide immediate protection to a list of Endangered Migratory Species, at present 40 mammals, birds, reptiles and fish. Further, in a second list are placed a much greater number of animals, including whole taxonomic groups (such as the Anseriformes and Falconiformes) which have an unfavourable conservation status and which require international agreements for their conservation and management. Guidelines for such agreements are set out in detail. The eventual network of agreements will be the real substance of the Convention – at present said by some to be the very skeleton of an umbrella. The working of the Convention will depend on an efficient Secretariat, to be provided for the first four years by the United Nations Environmental Programme (UNEP), and on the deliberations of a Scientific Council of the Parties.

The definition of 'Migratory Species' in the Bonn Convention is necessarily political in nature, being 'the entire population or any geographically separate part of the population of any species or lower taxon of wild animals, a significant proportion of whose members cyclically and predictably cross one or more boundaries defining national conservation and management jurisdiction for such species, as established in accordance with international law.'

An earlier Convention that also gives aid to migratory animals, though by no means wholly devoted to them, is the 'Convention on International Trade in Endangered Species of Wild Fauna and Flora' concluded at Washington, DC in 1973. As its name implies, the Convention seeks to control and reduce commerce in wild animals, their parts and products. Undoubtedly commercial greed has been a major factor in massive reductions in the numbers of, for instance, marine turtles and elephants, which would fit under the Bonn definition of migratory species. A weakness of the Washington Convention is that, although 50 countries have so far adhered to it, as long as one major exporting nation remains outside, it can serve as a convenient cover for all sorts of otherwise illegal trade. It is therefore of particular importance that this is the first international Convention concerned with wildlife that the Peoples' Republic of China has agreed to join. The Chinese have been notable for an insatiable demand for all sorts of odd animals, and their parts, in the pursuit of health and virility.

The Bonn Convention does make provision for habitat conservation, requiring in its guidelines for regional agreements 'Conservation, and where required and feasible, restoration of the habitats of importance in maintaining a favourable conservation status, and protection of such habitats from disturbance, including strict control of the introduction of, or control of already introduced, exotic species detrimental to migratory species' and, further, 'maintenance of a network of suitable habitats appropriately dispersed in relation to the migration routes' and 'where it appears desirable, the provision of new habitats . . . '. But these are at present just hopes and there is no doubt that nationalistic resistance is strongest where there is any attempt to get a state to modify its land-use policy.

It is therefore all the more commendable that the first of the global Conventions drawn up in the past decade was the 'Convention on Wetlands of International Importance especially as Waterfowl Habitat.' This was agreed in 1971 and is named the Ramsar Convention, not, as it might be thought, an acronym, but after the little town on the Caspian Coast of Iran where the concluding conference was held. It is now adhered to by 28 countries, including most of Europe and also Australia, South Africa, Iran, Pakistan, Soviet Union, Jordan, and Senegal. The United States and Canada are believed to be near signature, leaving then South America as the only continent where no country has as yet shown interest.

The Ramsar Convention is unusual in that it requires each Party to designate on an international List at least one of its major wetlands of international importance, precisely described and delimited on maps. Wetlands on this List must be maintained as far as possible in their present state, free of pollution and of human interference. The other Parties form the jury which decides whether the obligation is being met. So far 191 wetlands covering nearly six million hectares have been set aside from development in this way.

In practice no nation would accept an absolute embargo on its actions in perpetuity. So a let-out clause was necessary and it is possible 'because of its urgent national interests to delete or restrict the boundaries of wetlands already on the List'. However, if this is done the country 'should as far as possible compensate for any loss of wetland resources, and in particular it should create additional nature reserves for waterfowl and for the protection, either in the same area or elsewhere, of an adequate portion of the original habitat'. This compensation clause is, in fact, a tough one and serves as an effective block on any casual changes in designated wetlands.

In addition, the Parties undertake the wise use in general of wetlands within their territories, and to encourage necessary research and education.

The phrase 'especially as Waterfowl Habitat' was not, as might be thought, the brainchild of an arrogant ornithologist. It was proposed during the negotiations to indicate some limitation of the Convention's scope, and insisted upon by a major country. In fact it has not caused any notable restriction. But it does serve to recall the value of waterfowl (broadly defined in the Convention as birds ecologically dependent on wetlands) in deciding which wetlands are of international importance. One of the criteria is that the wetland should support at least one per cent of a regional population of a waterfowl species at some part of the year. Waterfowl are relatively easy to census, compared with many other wetland animals, and by their migratory movements the same birds confer importance on a chain of wetlands. It would be highly desirable if broader-based ecological criteria could be used, but the necessary surveys would, apart from considerations of manpower and cost, take so long that the wetland could well have been drained or otherwise destroyed before the results were available.

It is the vulnerability of wetlands to modern technology and the speed with which they can be destroyed that promoted the evolution of this special wetland Convention. We have already done our worst in Europe, but it is in the Third World Countries that the major ecological changes are about to be perpetrated such as the Jongli Canal scheme in Sudan and restructuring of the river basins in West Africa. We have had experience of how distant natural disasters, such as drought in the far-away Sahel, can severely reduce the numbers of insectivorous songbirds in Britain. We may soon be witnessing the impact of man-made disasters in the far south.

Perhaps the best hope for the future is in the World Conservation Strategy. This has been agreed by the United Nations' Environmental Programme (UNEP), the Food & Agricultural Organisation (FAO) and the International Union for the Conservation of Nature and Natural Resources (IUCN). It seeks a rational integrated approach to the problems of conservation. While recognizing the legitimate aspirations of the Third World, it endeavours to show how these may be met without destroying the ecology of our planet. If the nations of the world can be brought to accept the Strategy, all will not be lost. Otherwise our successors may not even know what a migrant animal is, let alone study the how and why of migrations.

R.ROBIN BAKER

Man and other vertebrates: a common perspective to migration and navigation

As the study of animal migration gathered momentum during the course of the twentieth century, there developed within it a deep-rooted parochialism. Students of the different animal groups, including Man, constructed terminological and conceptual frameworks that rendered any exchange of ideas difficult or even impossible. The resulting loss of perspective was eventually to the detriment of the entire subject.

The last decade has seen several attempts to rectify this situation, notably within the fields of orientation and navigation (Adler, 1971; Galler, Schmidt–Koenig, Jacobs & Belleville, 1972; Schmidt–Koenig & Keeton, 1978). The general study of migration has been much slower to seek the same perspective, though this present collection of papers bears testimony to the efforts that are now being made. Even in this volume, however, the legacy of the years of parochialism is still to be seen. Many different definitions of migration are offered and these still divide neatly, and contradictorily, between vertebrate-oriented definitions, involving periodic return movements, and invertebrate-oriented definitions not involving such return.

Despite containing such a relic of the parochial past, the intent of this collection of papers is to enable the reader automatically to reap the benefits of the improved perspective that comes from a comparative approach. In keeping with this intent, the aim of this final paper is to take the comparative approach to one of its extremes and consider whether there is anything to be learned about the migration patterns of humans from a study of the migration behaviour of other vertebrates. At the same time it considers the converse question of whether there is anything to be learned about the migration behaviour of other animals from a study of the migration behaviour of humans.

Inevitably, such a consideration runs the risk of lapsing into the ideological debate that has so bedevilled the scientific development of sociobiology since Wilson's original synthesis (Wilson, 1975). Usually, the sociobiology controversy (i.e. whether there is any justification for extrapolating from an understanding of the behaviour of other animals to an interpretation of the behaviour of Man) is debated on the battlefields of ideology and philosophy, even politics and dogma. In this paper such arenas are avoided. Instead, the theme is one of pragmatism; of practical benefit. Using three aspects of migration, two of them conceptual, the other experimental, this paper attempts to show that studies of the migration patterns of both Man and other vertebrates benefit considerably from the interchange of ideas that come from viewing each in the perspective of the other.

Year's home range, familiar area and exploration in vertebrates: the human scenario

Fig. 1 is a crude attempt to categorize the movement patterns of vertebrates, though in reality, of course, there is every gradation between the three situations shown. Categorization is solely for convenience. The solid black areas represent a **year's home range** (i.e. all the sites and routes used by an individual during the course of a year). Variation in the width of the lines making up the home range is intended to show that some parts are visited or travelled more often than others. No scale of distance is shown, being irrelevant in the present context. Details of the movement patterns of the animals listed below may be found in Baker (1978a).

Fig. 1(a) shows a pattern in which there is no seasonal shift in home range. Most often, there is some central roost or feeding site, etc. to which the animal

Fig. 1. Three types of year's home range in adult vertebrates. Black, all sites visited and routes used by an adult during the course of a year; dashed line, limits of familiar area.

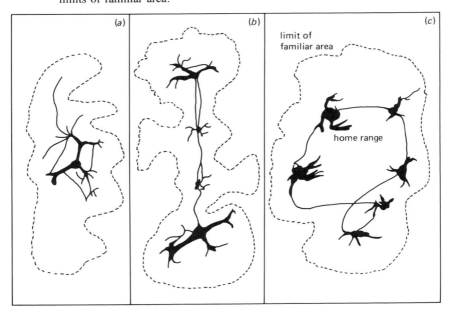

returns nearly every day of the year. This pattern is characteristic of a wide range of vertebrates such as coral fish (e.g. Pomacentridae), most terrestrial reptiles (e.g. lizards, snakes and tortoises), most small terrestrial mammals (e.g. rodents and insectivores), as well as many larger ones (e.g. bears and cats), most tropical bats (e.g. common vampire, *Desmodus rotundus*) and many birds (e.g. house sparrow, *Passer domesticus,* and feral pigeon, *Columba livia*).

Fig. 1(*b*) illustrates a regular to-and-fro shift between two seasonal home ranges (e.g. summer/winter; breeding/feeding; wet/dry). If these seasonal home ranges are far enough apart, relative to the animal's mobility, the individual may also use a succession of transient or 'stopover' home ranges during the migration from one range to another. Such a pattern is found in many stream-living and coastal fish (e.g. trout and some blennies), perhaps most amphibians (e.g. anurans and urodeles), some terrestrial reptiles (e.g. some tortoises, terrapins and snakes), all of the so-called 'migrants' among birds (e.g. barn swallow, *Hirundo rustica*), a few large terrestrial mammals (e.g. various antelope and deer), most temperate bats (e.g. American free-tailed bat, *Tadarida brasiliensis*), and nearly all baleen whales (e.g. blue whale, *Balaenoptera musculus*).

In Fig. 1(*c*) there are more than two seasonal home ranges and these are joined together to form an annual migration circuit. The pattern is probably typical of most oceanic vertebrates such as fish (e.g. herring, *Clupea harengus,* and cod, *Gadus morhua*), reptiles (e.g. sea turtles and sea snakes), birds (e.g. albatrosses and shearwaters) and many whales (e.g. perhaps most odontocetes) and on land some ungulates (e.g. wildebeest, *Connochaetus taurinus,* and saiga, *Saiga tatarica*), perhaps the males of many temperate bats (e.g. Indiana bat, *Myotis sodalis*) and perhaps some birds that live in tropical grassland and semi-desert habitats (e.g. red-billed quelea, *Quelea quelea,* and budgerigar, *Melopsittacus undulatus*), though in the latter areas the circuits may vary considerably from year to year.

In addition to the vertebrates already listed, Fig. 1(*a*)–(*c*) can also describe the year's home ranges of different groups of humans. Fig. 1(*a*), for example, will be an adequate description of the home range of not only the author and the reader but of almost every individual industrialist and agriculturalist. The vast majority of these humans have a life-style characterized by a fixed sleeping site to which the individual returns most nights. Fig. 1(*b*), on the other hand, describes the year's home range of those pastoral nomads (e.g. some Lapplanders and Fulani) that migrate to-and-fro with their ungulate herds between two major seasonal home ranges. Finally, Fig. 1(*c*) describes the home range of perhaps most hunter-gatherers (e.g. central-Australian aborigines and Kalahari bushmen) and many other pastoral nomads (e.g. Bedouin), particularly those that live in the great arid zone that stretches from the Atlantic Coast of North Africa eastwards to Tibet and western Manchuria.

Most forms of year's home range found in other vertebrates have their counterpart in one deme or other of humans. In our search for perspective, therefore, we can consider whether the insight that we must have into the nature of the **lifetime track** (i.e. path through time and space from birth to death) of the members of our own species can aid us in our attempt to understand the nature of the lifetime tracks of other species.

Consider the industrialist's home range shown in Fig. 1(a). It results entirely from familiarity with the area concerned, a knowledge of where and when to go to obtain or do this or that. Each individual's **familiar area** is the basis of that person's entire way of life. Yet a moment's thought will bring home the fact that familiarity extends to an area (shown by a dashed line in Fig. 1) much larger than the year's home range. Many parts of this larger area are either no longer visited or are visited infrequently. Usually, of course, there are good reasons why some parts of the familiar area continue to be used whereas others are abandoned. Most often, the areas used are in some way better, or more convenient, than those rejected. In effect, therefore, the year's home range has crystallized out from the much larger familiar area, a crystallization brought about by a process of habitat assessment and reassessment, adoption and rejection.

How do we humans build up our familiar area? First, we take as much advantage as possible of social communication, either by word of mouth or from books and magazines, etc. However, no such indirect method for learning about places is a good substitute for direct experience. By far the most important of all mechanisms for building up the familiar area is **exploration,** either solitarily or in the company of parents, other adults or contemporaries.

Exploration involves visiting a wide range of places, finding out what they have to offer, making comparisons, and finally rejecting some places and routes and adopting others. Comparisons may be made across spans of time as well as space. A site that offers little at one time of year may, at some other season, be superior to all others. Even industrialists show shifts in their monthly home range, though usually all are centred on the same sleeping place. In other cultures, where the relative suitability of the different places within an individual's familiar area changes much more drastically with the seasons, it may be necessary to shift the sleeping site seasonally, or perhaps even daily. Hence, for pastoral nomads and hunter-gatherers, the year's home range that crystallizes out from their familiar area is different from that for industrialists and agriculturalists. It follows that differences in the home ranges of humans, as shown in Fig. 1, can be attributed primarily to differences in the habitats that they encounter during exploration, rather than to any fundamental difference in the people themselves. Confronted by the stable, aseasonal habitat of a city, hunter-gatherers such as Australian aborigines or pastoral nomads such as Lapplanders are as likely to opt for a home range of the type shown in Fig. 1*a* as they are for a more traditional type of home range as in Fig. 1*b* or *c*.

Consideration of the lifetime tracks of humans, therefore, suggests that the different patterns of home range do not reflect fundamentally different mechanisms of establishment. All are produced by crystallization following an essentially similar process of familiarization by exploration and habitat assessment. It follows that the key to an understanding of the year's home range of adults lies in the programme for exploration by which the individual first builds up its familiar area and comes into contact with the range of habitats that will determine the final form of the home range. There appears to be no difference in the mechanisms used by so-called 'migrants' and 'non-migrants' save perhaps for differences in exploration programme. The human perspective, therefore, alerts us to the possibility that the same may be true for other vertebrates. This possibility has been discussed at length (Baker, 1978*a, b,* 1981*a, b*) and at least one study has shown that it is true for a bird. Over the past few decades, adults of the British breeding deme of the lesser black-backed gull, *Larus fuscus,* have effectively changed from being 'migrants' to being 'non-migrants'. Analysis of ringing returns, however, has shown that the pattern of exploration of young birds in their first two years has remained unchanged (Baker, 1980), continuing to involve the entire western seaboard of Europe from Britain south to northwest Africa. Whereas formerly, however, the birds when adult opted to winter in northwest Africa and Iberia, the majority now opt to winter in Britain.

Programmed restlessness: the innate orchestration of exploration

The human perspective seems to favour abandonment of the classical division of vertebrates into migrants and non-migrants. Instead it suggests that an understanding of the lifetime tracks of vertebrates lies with an understanding of the processes of exploration and habitat assessment. It is the exploration programme that determines the range of habitats an individual will encounter and the process of assessment that lies at the heart of the final crystallization of the adult home range.

When an individual is born, its familiar area is blank. By the age of first reproduction, we should expect the final home range more or less to have crystallized so that the individual is efficiently exploiting its environment. It follows that most exploration should be during the pre-reproductive phase of an individual's life and in most vertebrates this exploratory phase is easily recognized (Baker, 1978*a,* 1981*b*).

Fig. 2 shows the calculated relationship between the size of the home range for a variety of mammals, including Man, and the length in weeks of the pre-reproductive phase of exploration. Mammals that, as adults, live within a large home range have a proportionately longer phase of exploration than mammals that as adults live within a small home range. A similar analysis has not been carried out for birds. Nevertheless, insofar as those species with the largest of all home ranges (i.e. oceanic sea-birds which may, as adults, exploit entire oceans) often take several years to complete their explorations and begin to breed, a similar conclusion may be reached.

Such a correlation between size of adult home range and duration of exploration gives no proof of cause and effect. The tempting implication, however, is that species adapted to widely scattered resources as reflected by a large home range, need optimally to establish a relatively large familiar area before they begin to breed. In consequence, selection has favoured individuals with a relatively long phase of exploration. All this implies that the exploration programme is primarily the result of genetic evolution and that vertebrates are innately predisposed to carry out species-specific patterns of exploration. What evidence is there?

First, we need to be able to measure the incidence of exploration during an animal's lifetime. At present, this is virtually impossible in the field. In order to monitor the frequency with which an individual makes forays beyond the current limits of its familiar area, it is necessary to know those limits. Only by continuous radio-tracking of an individual from birth would this be possible, though much can be inferred about the exploration process from an analysis of the circumstantial information provided by intermittent field observations and the recovery of marked individuals (Baker, 1978a, chapter 17). As yet, however, there is no entirely satisfactory field measure of exploration and it has been suggested (Baker, 1978a) that much of the restlessness shown by animals in captivity is a specific reflection of the predisposition to explore and can be used in

Fig. 2. Relationship between final size of adult home range and duration of pre-reproductive phase of exploration in some mammals. Continuous line, primates (10 species); dashed line, carnivores, rodents, insectivores (13 species). Symbols indicate approximate positions for male and female humans. Simplified from an analysis in Baker (1978a, pp. 392–5) in which full details and raw data are presented. Both regression lines were a significant fit to the data at the 5% level.

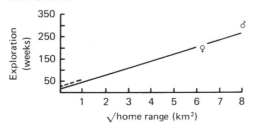

experimental studies. Two types of behaviour in particular seem to fall into this category. These are the so-called 'migratory restlessness' (*Zugunruhe*) of caged birds and the wheel-running behaviour of caged mammals.

Detailed arguments for interpreting 'migratory restlessness' of caged birds as reflecting a predisposition to explore rather than a predisposition to migrate in the classical sense have been presented elsewhere (Baker, 1981*b*). These arguments rest on the facts that: (1) many birds that do not 'migrate' in the classical sense nevertheless show 'migratory restlessness'; (2) 'migratory restlessness' in spring extends well beyond the normal seasonal migration period if the birds are deprived of certain resources; and (3) the ontogenetic sequence of restlessness is consistent with exploration. If we accept this point of view, then the recent work on restlessness (see Berthold, 1978; Gwinner & Czeschlik, 1978) becomes relevant to the subject under discussion. There is no doubt from this work that the duration and temporal pattern of restlessness, and hence exploration, is orchestrated by an innate programme. This programme, at least in some species of European warblers, is linked with an endogenous circannual clock, though it is likely normally to be phased and synchronized to the seasons by such exogenous factors as day-length and temperature. Timing and duration are not the only features of exploration that are under the influence of an innate endogenous programme. Some species (e.g. chaffinch, *Fringilla coelebs*, and garden warbler, *Sylvia borin*) have also been shown to have an innate preference for particular compass directions during different phases of restlessness (Perdeck & Clason, 1974; Gwinner & Wiltschko, 1978). It seems, then, that not only the timing and duration but also the direction of exploration may be orchestrated by innate programmes of variation in predispositions and thresholds.

The first suggestion that the wheel-running behaviour of caged mammals might be a specific reflection of an urge to explore rather than a general reflection of activity, boredom or a need for exercise was made in 1978 (Baker, 1978*a*). Since then, the suggestion has been considered in great detail (Mather, 1981*a*), both experimentally and in the light of an exhaustive literature review. Mather prefers to interpret the behaviour as reflecting an urge to reach unattainable resources. She suggests, however, that in an undisturbed laboratory situation this redirected activity is effectively a specific monitor of the urge to explore.

Fig. 3 shows ontogenetic curves of predisposition to explore in three species of mammal. All three show the pattern that has previously (Baker, 1978*a*) been argued to be characteristic of most vertebrates in that a peak is reached some time before first reproduction. In mammals, the peak usually comes during the subadult period, between the beginning of gametogenesis and the usual age of first reproduction or perhaps physical maturity. Having reached a peak, the incidence of exploration may plateau for a period, depending on how rapidly the individual encounters its various required resources, particularly suitable mates.

All of the examples in Fig. 3 are of individuals that fail to find a mate. Even in these, however, time spent in exploration eventually begins to decrease.

Whether the decrease in predisposition to explore in later life is part of an innate programme or reflects a constant predisposition interacting with the individual's familiarity with its environment is unresolved. There are strong indications, however, that the increase to a peak during the pre-reproductive period is part of such a programme. First, breeding experiments on rats and mice have shown that it is possible to produce lineages that spend greater or lesser amounts of time running in a wheel. Secondly, castration experiments (Fig. 3b) on rats

Fig. 3. Variation in incidence of exploration with age in three species of mammal. Age scale is years in (a) and days in (b) and (c). (a) Man: number of unmarried migrants (1970–1) into or out of the northwest region of Great Britain (to or from elsewhere in Great Britain) as a percentage of the total population in the northwest. Although not as specific a measure of exploration as (b) or (c), a large proportion of inter–regional migrants are likely to be exploratory (see arguments in Baker 1978a, chapter 14). Data extracted from 1971 Census of Great Britain by Anne Cronin (personal communication). (b) Laboratory rat: number of revolutions of running wheel per day of life. Castration and spaying data based on 8 males and 7 females for which lines are so similar that only one is shown (dotted line). Other data are 10-day means for 19 males (continuous line) and 19 females (dashed line). (Modified from Richter, 1933.) (c) Syrian hamster, *Mesocricetus auratus:* time spent (min) in running wheel per day of life (10-day means for 2 males continuous line and 2 females dashed line). (Data from Janice Mather, 1981b and personal communication.)

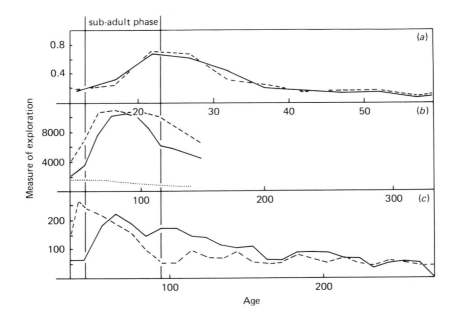

have shown that in the absence of a normal hormonal environment, the usual pattern of wheel-running does not develop but remains at a low level. A similar influence of castration on *exploration* will be familiar to anybody that owns a cat. Normal wheel-running performance of castrated rats can be restored by gonadal implants (Richter, 1933). There is an indication, therefore, that the normal progress of predisposition to explore in mammals is under the influence of an innate, hormonally mediated, programme.

What, then, of Man, showing as he does (Fig. 3a) the typical mammalian pattern for exploration? The implication is that the restlessness, the urge to travel and see new places, that we all associate with adolescence, is in reality a reflection of an innately programmed, hormonally mediated, urge to explore. The point cannot, of course, be proved to the satisfaction of the many people that resist the view that Man may be genetically predisposed to behave in particular ways (e.g. Gould, 1978). Nevertheless, there is an inference that the ontogenetic course of exploration in humans is under the control of an innate programme and is part of Man's vertebrate and mammalian inheritance.

Navigation and the avian lead to a new human sense

Fig. 4 is a stylized representation of a single exploratory foray beyond the current limits of an individual's familiar area. The usual function of such a foray will be to discover the sites of resources that are in short supply in the current familiar area. When a suitable new location is found an animal's immediate problem will be to incorporate the new site into its familiar area. Unless the outward exploratory route was a straight line, which is perhaps rarely the case, the animal has the task of pioneering the straightest, or at least the most economical, route between the new site and some point within its current home range (e.g. dotted line, Fig. 4). How does the animal do this? The process is that of **navigation** or goal-orientation.

Fig. 4. Representation of a single exploratory foray or experimental displacement beyond the limits of the present familiar area. Dotted line shows the most economical route back to the individual's home range (HR). (Redrawn from Baker, 1981a.)

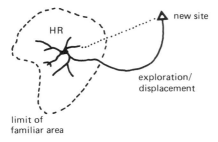

It is clear from Fig. 4 that navigation is an integral part of exploration. One other feature also emerges from Fig. 4. The process of experimental displacement during homing experiments (see Keeton, this volume) has much in common with the process of exploration. In both, the animal finds itself at a previously unfamiliar site and is faced with the task of pioneering a route back to its home range. The major difference is that during homing experiments animals are usually deprived of cues during the outward journey that would normally be available to them. Nevertheless, in principle, experimental displacement can be thought of as enforced exploration (Baker, 1978a, chapter 33).

If navigation is an integral part of exploration, it follows that all animals that explore require navigational ability and should be amenable to experimentation using displacement/release techniques. Man explores. There is no reason, therefore, to exclude Man from such experiments and the last few years have seen the beginnings of the study of human navigation using the techniques developed and honed over several decades in the study of pigeon navigation. The present state of the subject has been summarized (Baker, 1981a) but the field is likely to be one of exponential growth over the next decade. Only one aspect of human navigation without instruments is considered here, that of **route-based navigation** (i.e. the ability to monitor the direction and distance of the outward journey and to perform a double integration to vector the direction of home from the release point).

The first experimental series to study route-based navigation by humans was carried out in the Manchester area, England (Baker, 1981a). 'Home' was Manchester University. Small groups of between 5 and 11 University students were blindfolded and transported in a Sherpa van by a tortuous route from the University to release points at air-line distances of between 6 and 52 km from home (Fig. 5). Subjects were removed from the van individually and asked while still blindfold to state the compass direction of 'home' from the release point (i.e. N, NNE, NE, etc.). In Fig. 6 data are lumped for all releases in each of four compass quadrants relative to the University and show a strong ability of humans for route-based navigation even in the absence of visual cues.

On what senses and environmental information does this ability rest? Obvious clues such as the heat of the sun on one side of the face or familiarity with the route travelled, though they may influence the final estimate, are not the primary basis (Baker, 1981a). The accuracy of route-based navigation is not significantly reduced when the outward journey takes place under totally overcast skies nor does it improve significantly when subjects should be familiar with the later part of the outward journey. At the end of the first three years of study the sensory basis of route-based navigation remained as elusive as ever. The key to unravelling the mystery came from developments in the study of pigeon navigation.

There is a growing suspicion that pigeons are also capable of route-based navigation (Papi *et al.*, 1978; Kiepenheuer, 1978; Wiltschko, Wiltschko & Keeton, 1978; see Baker, 1981*b*, for review). Olfactory cues may be used (Papi *et al.*, 1978), but much more general seems to be the use of geomagnetic cues. Birds displaced in iron containers are more disoriented than birds displaced in aluminium containers (Papi *et al.*, 1978). Birds subjected to changed magnetic fields during the outward journey show deviations in their vanishing points when compared with controls (Kiepenheuer, 1978; Wiltschko *et al.*, 1978).

The first hint that the human ability for route-based navigation may also involve reference to the Earth's magnetic field came from an experiment carried out in June 1979 at Barnard Castle, Co. Durham on thirty-one sixth-form schoolchildren (16 to 17 years old). A group of 15 subjects wearing a small bar-magnet on the back of the head (tucked into the elastic of their blindfold) showed reduced accuracy of route-based navigation compared with a control group of 16 subjects wearing brass bars (Fig. 7). All subjects thought they were wearing magnets.

A second experimental series, carried out between October and November 1979, again based on Manchester University, was designed to test the possibility of a geomagnetic sense of direction in humans and its involvement in route-based navigation (Baker, 1981*a*). A group of 32 students was transported by coach on

Fig. 5. Release sites for the first experimental series on human navigation in the Manchester region. Hollow square, 'home' (Manchester University); filled triangles, A–H release points; stippled area, high ground. (Redrawn from Baker (1981*a*).)

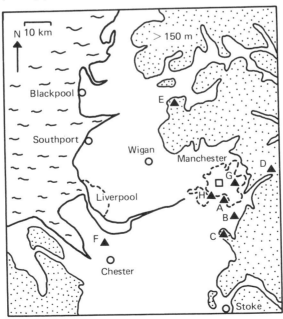

four separate journeys at weekly intervals to release points roughly south, west, north, and east of the University. As well as being blindfolded, each student wore a helmet supporting lateral copper coils (Fig. 8). The coils were connected to a 9-V battery carried by hand or in a pocket. All subjects were led to believe their helmets were activated but in reality 16 of the helmets contained a false wire and were inactive. These were the controls. Of the remaining 16 helmets, 8 (R-helmets) were such that the magnetic north pole was to the right of the head and 8 (L-helmets) to the left. Full experimental details are given in Baker (1981a).

At intervals during each outward journey (Fig. 9) the subjects were asked to write (while still blindfold) their estimate of the compass direction (N, NE, NNE, etc.) and air-line distance of the University from their current position. From this information each individual's estimate of the outward journey could be reconstructed.

Figs. 9 and 10 confirm that route-based navigation by humans involves reference to the Earth's magnetic field and suggests that the primary reference

Fig. 6. A demonstration of route-based navigation by humans. Each filled circle represents one individual's estimate (while still blindfold) of the compass direction of 'home' (drawn relative to true home direction). Arrows, mean vectors. All distributions are significantly non-uniform ($P < 0.05$) (Rayleigh test, Batschelet, 1965). (Simplified from Baker 1981a).)

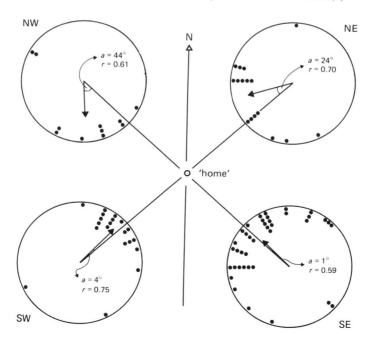

involves a geomagnetic sense of direction. Manipulation of the magnetic field passing through the head influences ability to estimate the direction of travel during the various legs of the outward journey.

When the experiment was designed it was naively assumed that the coils would create a fixed field through the head irrespective of the direction of travel. On this model, subjects wearing R-helmets should always assess themselves to be travelling west and those wearing L-helmets always to be travelling east, irrespective of the actual direction of travel. The data (Fig. 10), however, suggest that in the region of whatever sense organ is involved, the field created by the helmets interacts with, rather than overrides, the Earth's magnetic field. A plot of the horizontal component of the lines of force passing through the head (e.g. Fig. 11) shows that there is such interaction in the periphery of the head. So far, however, only one position has been found where the variation in horizontal component when facing in different compass directions matches the

Fig. 7. The Barnard Castle experiment: first indication of the involvement of the geomagnetic field in route-based navigation by humans. Each dot or cross indicates an individual's written estimate (while still blindfold) of the compass direction of 'home' from the two sites indicated (hollow squares). Filled circles, controls; crosses, individuals wearing a bar-magnet on the head. Mean vectors (arrows) are shown only if based on significantly ($P < 0.05$) non-uniform distributions (Z and P values). X and P values are statistics for Stephens' exact test (Batschelet, 1965) which compares observed mean angle with home direction. At both positions, the controls produced a good estimate of home direction. At the first position, the group wearing magnets produced an estimate significantly anticlockwise of the home direction whereas, at the second position, the estimate was not significantly non-uniform. (Redrawn from Baker (1981b).)

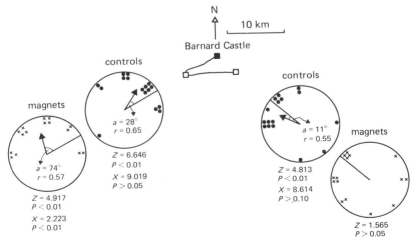

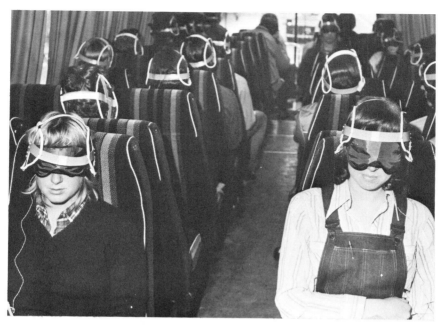

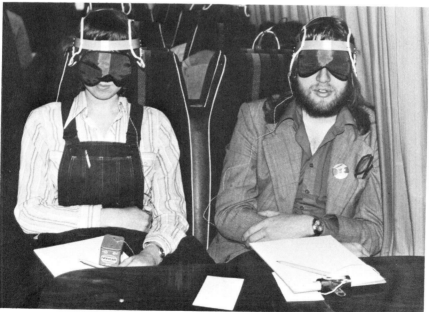

Fig. 8. Helmets used in the second Manchester experimental series on human navigation. Each helmet bears two lateral coils, each of 200 turns of 40 standard-wire-gauge copper. A 50-mA current from a 9 V battery passed

Fig. 9. Routes taken during second Manchester series and the influence of a changed magnetic field on route-based navigation in humans. Four lines show routes taken (weekly intervals). All journeys began at Manchester University (filled circle). Hollow triangles, positions at which subjects were asked to write (while still blindfold) their estimate of the compass direction and air-line distance of the University. Data shown are the mean vectors (a°, r) of the estimates at the final position on each journey. All estimates by controls are within 30° of the true home direction and (except for day 4) are significantly non-uniform. Except on day 3, estimates by the group wearing magnets are not significantly non-uniform. On day 2, mean angles for the R- and L-helmet groups were significantly different $(P < 0.05)$. (For raw data see Baker (1981a).)

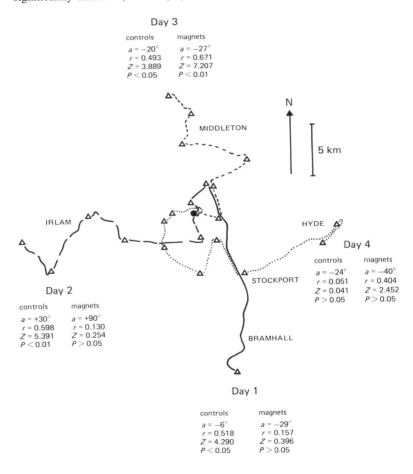

Day 3

controls	magnets
$a = -20^\circ$	$a = -27^\circ$
$r = 0.493$	$r = 0.671$
$Z = 3.889$	$Z = 7.207$
$P < 0.05$	$P < 0.01$

N

5 km

MIDDLETON

IRLAM

HYDE

Day 4

controls	magnets
$a = -24^\circ$	$a = -40^\circ$
$r = 0.051$	$r = 0.404$
$Z = 0.041$	$Z = 2.452$
$P > 0.05$	$P > 0.05$

STOCKPORT

Day 2

controls	magnets
$a = +30^\circ$	$a = +90^\circ$
$r = 0.598$	$r = 0.130$
$Z = 5.391$	$Z = 0.254$
$P < 0.01$	$P > 0.05$

BRAMHALL

Day 1

controls	magnets
$a = -6^\circ$	$a = -29^\circ$
$r = 0.518$	$r = 0.157$
$Z = 4.290$	$Z = 0.396$
$P < 0.05$	$P > 0.05$

Caption to Fig. 8 (cont.).
through both coils produced a magnetic field (see Fig. 11) at the centre of the helmet 3.5 times the horizontal component of the Earth's field. For further experimental details see Baker (1981a). (Photographs by Les Lockey.)

observed variation in estimates of the direction of travel shown in Fig. 10 (Baker, 1981a). This position is a point midway between and slightly below the eyes about 3 cm in from a line dropped vertically from the leading edge of the helmet. It is suggested (Baker, 1981a) that this position is the location of the magnetic sense organ in humans. In pigeons, a candidate for the magnetic sense organ, consisting of particles of magnetite, has been described (Walcott, Gould & Kirschvink, 1979). This organ is unpaired and is situated between the dura mater and the skull, though the precise location has not yet been described.

Discussion

The human perspective provides a strong conceptual framework within which to study the lifetime tracks of vertebrates. The picture produced is that of exploration and familiarization along with, and followed by, habitat assessment

Fig. 10. Effect of a changed magnetic field on the estimate of direction of travel by blindfold humans. When travelling northwest ($\pm 45°$), there is no difference in estimate of direction of travel between controls and the two experimental groups. When travelling northeast ($\pm 45°$), there is no difference between the two experimental groups but the combined estimate is significantly different from that of controls. When travelling south ($\pm 90°$), R- and L-helmet wearers produce significantly different estimates of direction of travel either side of the estimate made by controls. The R-helmet estimate is also significantly different from that made by controls (Watson & Wheeler parametric two-sample test: Batschelet, 1965). (For raw data and further statistics see Baker (1981a).)

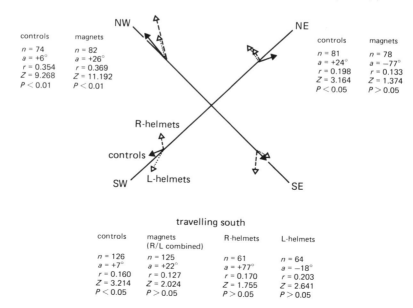

controls	magnets		controls	magnets
$n = 74$	$n = 82$		$n = 81$	$n = 78$
$a = +6°$	$a = +26°$		$a = +24°$	$a = -77°$
$r = 0.354$	$r = 0.369$		$r = 0.198$	$r = 0.133$
$Z = 9.268$	$Z = 11.192$		$Z = 3.164$	$Z = 1.374$
$P < 0.01$	$P < 0.01$		$P < 0.05$	$P > 0.05$

travelling south

controls	magnets (R/L combined)	R-helmets	L-helmets
$n = 126$	$n = 125$	$n = 61$	$n = 64$
$a = +7°$	$a = +22°$	$a = +77°$	$a = -18°$
$r = 0.160$	$r = 0.127$	$r = 0.170$	$r = 0.203$
$Z = 3.214$	$Z = 2.024$	$Z = 1.755$	$Z = 2.641$
$P < 0.05$	$P > 0.05$	$P > 0.05$	$P > 0.05$

and crystallization of the adult home range. Perhaps eventually it will be found that only by the adoption of this scenario will students of the different vertebrate groups achieve realistic models of the lifetime tracks of the particular animals in which they are interested.

Acceptance of the same scenario for human and non-human vertebrates alike invites us in our attempts better to understand the human condition to extrapolate from experiments, such as breeding and castration studies, that are done only on animals other than man. Such extrapolation, as in the example of programmed exploration presented in this paper, can give us insight into human behaviour patterns that otherwise would be impossible to obtain. The opponents of socio-biology are unlikely to be enamoured by this extrapolation and will undoubtedly continue to take refuge behind the impossibility, without direct experimentation, of obtaining critical data on whether humans are innately predisposed to behave

Fig. 11. Magnetic fields produced by R- and L-helmets as a function of direction of travel. The figures show plan views through R- and L-helmets at a plane about eyebrow level (see Fig. 8). Outer oblong, outline of perimeter of helmet. Solid circles, positions of coils; dashed line in upper-left figure, approximate outline of brain; arrowheads, polarity of horizontal vector (arrow pointing toward magnetic north); x, null point in the horizontal field.

In the centre of the head the field changes little as the helmet wearer faces different directions. At the periphery, the artificial field interacts noticeably with the Earth's field. At the plane shown, no position in the head matches the observed variation in estimate shown in Fig. 10. At a slightly lower plane a position emerges as described in the text. (Redrawn from a plot by S. E. R. Bailey, personal communication.)

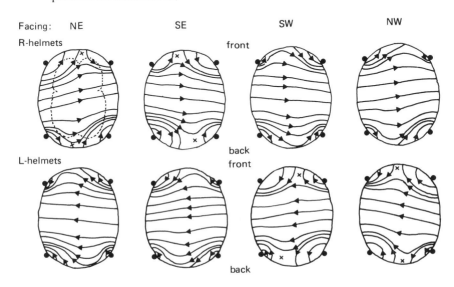

in particular ways. Many others, however, may consider the results of such extrapolation to be informative and acceptable, given the ethical impossibility of carrying out on humans the crucial experiments.

The final benefit from using the same conceptual framework for humans and non-humans comes when, as in the case of navigation, experiments on other animals give a strong indication of a profitable line of research on Man. The author, for one, would certainly never have conceived, let alone acted upon, the potential of blindfolding fellow humans, placing a magnet on their heads, and transporting them to some unfamiliar site, had it not been for the recent developments in the study of pigeon navigation combined with acceptance of the approach outlined in this paper.

Philosophical and theoretical debate on the validity of adopting a common scenario for the lifetime tracks of Man and other vertebrates will doubtless continue for some time. The practical benefits of the resulting perspective, however, seem already to be showing themselves.

Summary

(1) The study of animal migration is in the process of emerging from a phase of parochialism.

(2) Most forms of the year's home range found in other vertebrates have their counterpart in one deme or other of humans. In humans, the year's home range crystallizes out from a much larger familiar area by a process of habitat assessment and reassessment, adoption and rejection. The most important of all mechanisms for building up the familiar area is exploration. The human perspective seems to favour an abandonment of the classical division of vertebrates into 'migrants' and 'non-migrants'. Instead, it suggests that an understanding of the lifetime tracks of vertebrates lies with an understanding of the processes of exploration and habitat assessment.

(3) Most exploration occurs during the pre-reproductive phase of an animal's life. The duration of the exploratory phase is a function of the size of the home range the animal is likely to occupy when adult. There is evidence for birds and mammals that the duration, temporal pattern and, at least in some birds, compass direction of exploration is orchestrated by an innate, hormonally mediated, programme of variation in predispositions and thresholds. Man shows a typical vertebrate pattern of exploration during ontogeny. The implication is that the restlessness, the urge to travel and see new places, that is associated with adolescence in humans, is also a reflection of an innately programmed, hormonally mediated, urge to explore.

(4) Navigation is an integral part of exploration. Moreover, in principle,

experimental displacement during homing experiments can be thought of as enforced exploration. All animals that explore, including Man, need to be able to navigate and should be amenable to experimentation by displacement-release. Man shows a strong ability for route-based navigation even when blindfold during the outward journey. There is a growing suspicion that the same is true for pigeons and that, if so, the ability is based on a geomagnetic sense of direction. This development in the study of pigeon navigation has led to the discovery of a similar use of a geomagnetic sense of direction during route-based navigation in humans.

(5) Philosophical debate on the justification for adopting a common scenario for the lifetime tracks of Man and other vertebrates will undoubtedly continue for some time. The practical benefits of the resulting perspective, however, seem already to be showing themselves.

References

Adler, H. E. (Ed.) (1971). *Orientation: Sensory Basis,* 188 pp. New York: N.Y. Academy of Science.

Baker, R. R. (1978*a*). *The Evolutionary Ecology of Animal Migration.* London: Hodder & Stoughton.

Baker, R. R. (1978*b*). Demystifying vertebrate migration. *New Scientist,* **80,** 526–8.

Baker, R. R. (1980). The significance of the lesser black-backed gull, *Larus fuscus,* to models of bird migration. *Bird Study,* **27,** 41–50.

Baker, R. R. (1981*a*). *Human Navigation and the Sixth Sense.* London: Hodder & Stoughton.

Baker, R. R. (1981*b*). *Migration: Paths Through Time and Space.* London: Hodder & Stoughton.

Batschelet, E. (1965). *Statistical Methods for the Analysis of Problems in Animal Orientation and Certain Biological Rhythms.* Washington, D.C.: The American Institute of Biological Sciences.

Berthold, P. (1978). Concept of endogenous control of migration in warblers. In *Animal migration, navigation and homing,* ed. K. Schmidt–Koenig & W. T. Keeton, pp. 295–82. Heidelberg: Springer.

Galler, S. R., Schmidt–Koenig, K., Jacobs, G. J. & Belleville, R. E. (eds.) (1972). *Animal Orientation and Navigation.* Washington, D.C.: Scientific and Technical Information Office, National Aeronautics and Space Administration.

Gould, S. J. (1978). Sociobiology: the art of storytelling. *New Scientist,* **80,** 530–3.

Gwinner, E. & Czeschlik, D. (1978). On the significance of spring migratory restlessness in caged birds. *Oikos,* **30,** 364–72.

Gwinner, E. & Wiltschko, W. (1978). Endogenously controlled changes in migratory direction of the garden warbler, *Sylvia borin. Journal of Comparative Physiology,* **125,** 267–73.

Kiepenheuer, J. (1978). Inversion of the magnetic field during transport: its influence on the homing behaviour of pigeons. In *Animal migration, Navi-*

gation and Homing, ed. K. Schmidt–Koenig & W. T. Keeton, pp. 135–42. Heidelberg: Springer.

Mather, J. G. (1981*a*). Wheel-running activity: a new interpretation. *Mammal Review,* **11** (1).

Mather, J. G. (1981*b*). *Wheel-running, Exploration and Navigation in Rodents.* Unpublished Ph.D. Thesis, University of Manchester.

Papi, F., Ioalé, P., Fiaschi, V., Benvenuti, S. & Baldaccini, N. E. (1978). Pigeon homing: cues detected during the outward journey influence initial orientation. In *Animal Migration, Navigation and Homing,* ed. K. Schmidt–Koenig & W. T. Keeton, pp. 65–77.

Perdeck, A. C. & Clason, C. (1974). Spontaneous migration activity of chaffinches in the Kramer cage. *Progress Report 1973, Institute of Ecological Research, Royal Netherlands Academy of Arts and Sciences,* pp. 81–2.

Richter, C. P. (1933). The effect of early gonadectomy on the gross body activity of rats. *Endocrinology,* **17,** 445–50.

Schmidt–Koenig, K. & Keeton, W. T. (eds.). (1978). *Animal Migration, Navigation and Homing.* Heidelberg: Springer.

Walcott, C., Gould, J. L. & Kirschvink, J. L. (1979) Pigeons have magnets. *Science,* **205,** 1027–8.

Wilson, E. O. (1975) *Sociobiology, the New Synthesis.* Harvard: Belknap Press.

Wiltschko, R., Wiltschko, W. & Keeton, W. T. (1978) Effect of outward journey in an altered magnetic field on the orientation of young homing pigeons. In *Animal Migration, Navigation and Homing,* ed. K. Schmidt–Koenig & W. T. Keeton, pp. 152–61. Heidelberg: Springer.

INDEX OF MIGRANTS

SUBJECT INDEX